AF767994

Mitigating Contamination from Food Processing

Food Chemistry, Function and Analysis

Series editors:
Gary Williamson, *University of Leeds, UK*
Alejandro G. Marangoni, *University of Guelph, Canada*
Juliet A. Gerrard, *University of Auckland, New Zealand*

Titles in the series:
1: Food Biosensors
2: Sensing Techniques for Food Safety and Quality Control
3: Edible Oil Structuring: Concepts, Methods and Applications
4: Food Irradiation Technologies: Concepts, Applications and Outcomes
5: Non-extractable Polyphenols and Carotenoids: Importance in Human Nutrition and Health
6: Cereal Grain-based Functional Foods: Carbohydrate and Phytochemical Components
7: Steviol Glycosides: Cultivation, Processing, Analysis and Applications in Food
8: Legumes: Nutritional Quality, Processing and Potential Health Benefits
9: Tomato Chemistry, Industrial Processing and Product Development
10: Food Contact Materials Analysis: Mass Spectrometry Techniques
11: Vitamin E: Chemistry and Nutritional Benefits
12: Anthocyanins from Natural Sources: Exploiting Targeted Delivery for Improved Health
13: Carotenoid Esters in Foods: Physical, Chemical and Biological Properties
14: Eggs as Functional Foods and Nutraceuticals for Human Health
15: Rapid Antibody-based Technologies in Food Analysis
16: DNA Techniques to Verify Food Authenticity: Applications in Food Fraud
17: Advanced Gas Chromatography in Food Analysis
18: Handbook of Food Structure Development
19: Mitigating Contamination from Food Processing

How to obtain future titles on publication:
A standing order plan is available for this series. A standing order will bring delivery of each new volume immediately on publication.

For further information please contact:
Book Sales Department, Royal Society of Chemistry, Thomas Graham House, Science Park, Milton Road, Cambridge, CB4 0WF, UK
Telephone: +44 (0)1223 420066, Fax: +44 (0)1223 420247
Email: booksales@rsc.org
Visit our website at www.rsc.org/books

Mitigating Contamination from Food Processing

Edited by

Catherine S. Birch
AgriFoodX Limited, Chester, UK
Email: cath.birch@agrifoodx.com

and

Graham A. Bonwick
AgriFoodX Limited, Chester, UK
Email: graham.bonwick@agrifoodx.com

ROYAL SOCIETY OF CHEMISTRY

Food Chemistry, Function and Analysis No.19

Print ISBN: 978-1-78262-922-1
PDF ISBN: 978-1-78801-643-8
EPUB ISBN: 978-1-78801-906-4
Print ISSN: 2398-0656
Electronic ISSN: 2398-0664

A catalogue record for this book is available from the British Library

© The Royal Society of Chemistry 2020

All rights reserved

Apart from fair dealing for the purposes of research for non-commercial purposes or for private study, criticism or review, as permitted under the Copyright, Designs and Patents Act 1988 and the Copyright and Related Rights Regulations 2003, this publication may not be reproduced, stored or transmitted, in any form or by any means, without the prior permission in writing of The Royal Society of Chemistry or the copyright owner, or in the case of reproduction in accordance with the terms of licences issued by the Copyright Licensing Agency in the UK, or in accordance with the terms of the licences issued by the appropriate Reproduction Rights Organization outside the UK. Enquiries concerning reproduction outside the terms stated here should be sent to The Royal Society of Chemistry at the address printed on this page.

Whilst this material has been produced with all due care, The Royal Society of Chemistry cannot be held responsible or liable for its accuracy and completeness, nor for any consequences arising from any errors or the use of the information contained in this publication. The publication of advertisements does not constitute any endorsement by The Royal Society of Chemistry or Authors of any products advertised. The views and opinions advanced by contributors do not necessarily reflect those of The Royal Society of Chemistry which shall not be liable for any resulting loss or damage arising as a result of reliance upon this material.

The Royal Society of Chemistry is a charity, registered in England and Wales, Number 207890, and a company incorporated in England by Royal Charter (Registered No. RC000524), registered office: Burlington House, Piccadilly, London W1J 0BA, UK, Telephone: +44 (0) 20 7437 8656.

Visit our website at www.rsc.org/books

Printed in the United Kingdom by CPI Group (UK) Ltd, Croydon, CR0 4YY, UK

Preface

It has been suggested that the discovery of fire and the subsequent development of cooking by the ancestors of modern humans assisted the evolution of speech and intelligence, by reducing the need for powerful jaw muscles to chew raw food and by enhancing the bioavailability of nutrients therein. Thermal and other food processing techniques have since enabled the diverse diet of microbiologically safe, palatable food that modern consumers expect. Although food processing provides numerous benefits, there is a growing awareness of the chemical-derived changes that can occur, especially during thermal processing. These changes give rise to a range of chemical contaminants, the consumption of which has been linked to diverse negative health effects. Given the increasing prevalence of processed foods within the modern diet, it is essential that steps are taken to limit consumer exposure to these contaminants through the understanding of their formation and the application of mitigation strategies. This book provides an updated review of food-derived process contaminants, starting with an overview of current European Union regulations that have been established to prevent or minimise consumer exposure (Chapter 1). Subsequent chapters explore the formation and mitigation strategies of well recognised priority contaminants such as acrylamide (Chapters 2 and 3) and the monochloropropanediols, glycidol and their esters (Chapters 4, 5 and 7). Of particular concern are those contaminants that occur in post-processed ingredients and thus foods that are either widely or frequently consumed. Vulnerable members of society such as new-borns, children or those with long-term health conditions may be particularly at risk from these process contaminants and their detrimental effects on wellbeing. Genotoxic contaminants are of particular concern, such as the polynuclear aromatic hydrocarbons, and these are addressed in Chapter 6. Although

Food Chemistry, Function and Analysis No.19
Mitigating Contamination from Food Processing
Edited by Catherine S. Birch and Graham A. Bonwick
© The Royal Society of Chemistry 2020
Published by the Royal Society of Chemistry, www.rsc.org

not currently considered to be a priority contaminant and not specifically regulated, the advanced glycation end products (AGEs), are discussed in Chapter 8. The diverse range of health impacts that are beginning to be associated with elevated levels of these AGE compounds and their contribution to increased risk of disease states linked to their consumption are comprehensively reviewed. For these reasons, it is likely that this group of contaminants will attract more interest, potentially as markers of 'over processed' foods and will thus be subject to regulation in the future.

This book is designed to provide an introduction to the area of process contaminants and an insight into recent research. It is expected that this will be of interest to students, researchers and regulators, as well as food business operators. The individual authors of each chapter aim to increase understanding of the formation of each of the process-derived contaminants and to highlight current identification and mitigation strategies. The overall aim is to increase awareness of these contaminants and to ensure that the increasing global population can continue to be fed with products of that are of high quality and safe for consumption.

Catherine S. Birch and Graham A. Bonwick

Biography

Prof. Graham Bonwick established AgriFoodX Limited with Dr Catherine Birch in July 2019. Prior to this he was Senior Lecturer and Lead Scientist in the Institute for Agri-Food Research & Innovation (a joint venture between Newcastle University and Fera Science Ltd), after serving as the Director of the Institute of Food Science and Innovation at the University of Chester. Graham has undertaken applied research and developed training to support growth and innovation of the food and drink sector industries. His research interests are environmental quality and food safety, and he is currently involved in activities relating to the circular bioeconomy, such as the characterisation and evaluation of agri-food by-products for bioactive compounds and bio-based packaging solutions. Graham holds Fellowships with the Royal Society of Biology and the Royal Society of the Arts and is also a member of the Institute of Food Science & Technology and the Institute of Biomedical Science.

Dr Catherine Birch is a Director of AgriFoodX Limited which she co-established in July 2019. Formerly a Research Fellow at Newcastle University she has experience of research into Functional Foods, Nutraceuticals and Bioactive Ingredients and Nanotechnology. Cath's personal research has involved successful development of novel compounds that help to ameliorate vascular disease. She has previously developed and taught food science and packaging programmes at both undergraduate and postgraduate level, as well as for industry up-skilling. Cath has extensive experience of business engagement and applied research to support innovation and growth, which has included the establishment of networks for artisan food and drink producers. She is a Member of the Royal Society of Biology, Fellow of the Higher Education Academy and Fellow of the Institute of Food Science & Technology and an enthusiastic science communicator.

Food Chemistry, Function and Analysis No.19
Mitigating Contamination from Food Processing
Edited by Catherine S. Birch and Graham A. Bonwick
© The Royal Society of Chemistry 2020
Published by the Royal Society of Chemistry, www.rsc.org

Contents

Food Chemistry, Function and Analysis No.19
Mitigating Contamination from Food Processing
Edited by Catherine S. Birch and Graham A. Bonwick

© The Royal Society of Chemistry 2020

Published by the Royal Society of Chemistry, www.rsc.org

CHAPTER 1

European Regulation of Process Contaminants in Food

GRAHAM A. BONWICK* AND CATHERINE S. BIRCH

AgriFoodX Limited, Chester, UK
*E-mail: graham.bonwick@agrifoodx.com

1.1 Background

Process contaminants are substances that form in food or in food ingredients when they undergo chemical changes during processing. Processing methods include heat treatment such as cooking, as well as fermentation, smoking, drying and refining. Many foods must be cooked to make them edible and digestible, or to improve their organoleptic properties and acceptability. However, thermal processing of foods by baking, frying, grilling or barbecuing, either at home or during manufacture, can have undesired consequences. Besides the loss of some nutrients, for example vitamins, potentially harmful by-products can also develop.

Within the European Union (EU), contaminants in food products were initially addressed by the framework regulation for food contaminants Council Regulation (EEC) No. 315/93 of 8th February 1993. For the purposes of the regulation, a contaminant may be defined as: '*any substance not intentionally added to food which is present in such food as a result of the production (including operations carried out in crop husbandry, animal husbandry*

Food Chemistry, Function and Analysis No.19
Mitigating Contamination from Food Processing
Edited by Catherine S. Birch and Graham A. Bonwick
© The Royal Society of Chemistry 2020
Published by the Royal Society of Chemistry, www.rsc.org

and veterinary medicine), manufacture, processing, preparation, treatment, packing, packaging, transport or holding of such food or as a result of environmental contamination'. To ensure public health, contaminant levels should be kept as low as can be reasonably achieved through good manufacturing, agricultural and fishery practices and which are toxicologically acceptable. Council Regulation (EEC) No. 315/93 stipulated that maximum levels were to be set for certain contaminants. Paragraph 3 of Article 2 of the regulations requires that maximum levels must be set for specific contaminants and that these levels must be adopted in the form of a non-exhaustive Community list. Community measures relating to specific contaminants were introduced by Commission Regulation (EC) No 466/2001 which was designed to address disparities between the laws of Member States in regard to the maximum levels for contaminants in certain foodstuffs and the consequent risk of distortion of competition. Maximum levels for lead, cadmium, mercury, dioxins, polycyclic aromatic hydrocarbons (PAHs), nitrate, 3-monochloropropanediol (3-MCPD), aflatoxins, ochratoxin A, patulin and inorganictin were set under this legislation.

Council Regulation (EC) No 178/2002 set down the general principles and requirements of food law within the EU, established the European Food Safety Authority (EFSA) and general procedures in matters of food safety. Subsequently Regulation (EC) No 882/2004 established the basis for official controls to be performed by EU member states to ensure the verification of compliance with feed and food law, animal health and animal welfare rules. Maximum levels of process contaminants were further addressed in Commission Regulation (EC) No. 1881/2006 and subsequent amendments. Commission Regulation (EC) No. 1881/2006 also stipulated that 'in the cases of contaminants which are considered genotoxic carcinogens or in cases where current exposure of the population or vulnerable groups in the population is close to or exceeds the tolerable intake, maximum levels should be set at a level which is as low as reasonably achievable.' The regulations reinforce the role of Food Business Operators (FBOs) as key contributors towards the protection of public health. Contaminated products exceeding the maximum permitted levels should not be placed on the market, nor mixed with other foodstuffs or used as an ingredient. The Commission Regulation (EC) No. 1881/2006 also specified sampling and analysis measures to achieve uniformity in the reporting and interpretation of contaminants by the competent authorities of the member states.

The following have been identified by the EU as the major process contaminants of concern in food to date and as such their occurrence and concentrations are expected to be minimised to ensure public health protection:

- Acrylamide
- Chloropropanols [3-monochloropropanediol (3-MCPD), 2-monochloropropanediol (2-MCPD)] and their esters

- Glycidyl fatty acid esters (GEs)
- Furan and methylfurans
- Polycyclic aromatic hydrocarbons (PAHs)
- Ethyl carbamate
- Heterocyclic amines
- Hydroxymethylfurfuryl (HMF)

1.2 Acrylamide

Acrylamide is mainly formed in food by the reaction of asparagine with reducing sugars (particularly glucose and fructose) as part of the Maillard reaction; acrylamide may also be formed *via* reactions involving 3-aminopropionamide.[1,2] The International Agency for Research on Cancer (IARC) classified acrylamide as probably carcinogenic to humans (group 2A) in 1994. Subsequently, the EFSA Scientific Opinion on acrylamide in food (2015) described acrylamide as a cause for concern for human health with young children as the most vulnerable group.[3]

Commission Regulation (EU) 2017/2158 came into effect in April 2018. This regulation established best practice and describes the mitigation measures required to minimise acrylamide formation during processing and to enable reduced concentrations in processed foods. The regulation also sets benchmark levels (BMLs; formerly referred to as Indicative Values) for processed foods. BMLs were established to provide guidance to FBOs such that consumer exposure to acrylamide would be reduced.

The food categories addressed by Commission Regulation (EU) 2017/2158 apply to all EU FBOs and include:

1. Products based on raw potatoes
   a. Food business operators (FBOs) to identify and use appropriate varieties
   b. Sliced potato crisps
   c. French fries and other cut deep fried or oven fried potato products
2. Dough-based potato crisps, snacks, crackers and other dough-based potato products
3. Fine bakery wares – cookies biscuits, rusks, cereal bars, scones, cornets, wafers, crumpets and gingerbread, crackers (dry biscuit based on cereal flour), crispbreads and bread substitutes
4. Breakfast cereals
5. Coffee: (i) roast coffee; (ii) instant (soluble) coffee
6. Coffee substitutes containing more than 50% cereals
7. Coffee substitutes containing more than 50% chicory
8. Baby biscuits and infant cereals
9. Baby jar foods [low acid and prune-based foods as per European Regulation (EU) 609/2013]
10. Bread

In addition to setting BMLs, Commission Regulation (EU) 2017/2158 addresses FBO mitigation measures, sampling and analysis procedures. FBOs within the EU are expected have a general understanding of acrylamide as a food safety hazard and the conditions leading to its formation. Consequently, FBOs are expected to undertake steps necessary to mitigate the formation of acrylamide in the food that they produce and to implement appropriate food safety management procedures such as those based on Hazard Analysis and Critical Control Point (HACCP) principles. Records should be kept of any sampling and analysis undertaken, as well as any mitigation measures implemented. The food product sampling plans and the results of testing should also be recorded. It is intended that the measures are proportionate to the size and complexity of the business and that small and micro FBOs are not unduly burdened and economically disadvantaged. To assist national and local authorities, as well as FBOs, FoodDrinkEurope makes available an Acrylamide 'Toolbox' to assist implementation of Commission Regulation (EU) 2017/2158 and reduce public exposure to acrylamide. This can be accessed at http://www.fooddrinkeurope.eu/publication/fooddrinkeurope-updates-industry-wide-acrylamide-toolbox/.

Acrylamide formation primarily takes place under conditions of high temperature (usually in excess of 120 °C) and low moisture, both in industrial food processing and in domestic cooking processes such as frying, baking and roasting.[1] For these reasons, cooking directions should be provided to consumers for home cooking which ensure microbiological safety, whilst minimising acrylamide formation. Recommended cooking methods specifying time, temperature, quantity for oven/deep fryer/pan on packaging and/or *via* other communication channels must be provided to end users. For consumers the recommended cooking instructions shall be clearly displayed on all product packaging in compliance with Regulation (EU) No 1169/2011 that addresses the provision of food information to consumers. In the United Kingdom, The Food Standards Agency advises consumers to 'aim for a golden yellow colour or lighter when frying, baking, toasting or roasting starchy foods' and to 'follow the cooking instructions on the pack when cooking packaged foods like chips and roast potatoes'.

1.3 3-Monochloropropanediol (3-MCPD), 2-MCPD and Their Esters, and Glycidyl Fatty Acid Esters (GEs)

Acid hydrolysed vegetable proteins are used as flavour enhancers and ingredients in processed savoury foods. Hydrolysis with hydrochloric acid (HCl) combined with poor time and/or temperature control and incorrect hydrolysate neutralisation procedures can generate 3-MCPD, 2-MCPD and related compounds.[4] At-risk foods include:

- Soy sauce
- Oyster sauce
- Broths
- Soups
- Savoury snacks
- Gravy mix
- Stock cubes

Replacement of HCl with sulphuric acid is performed to prevent production of chlorinated compounds but the fermentation products often require treatment with colourants (*e.g.* caramel) and flavour enhancers (*e.g.* monosodium glutamate, soy sauce). Consequently, additional food safety issues reported through the EU Rapid Alert System for Food Products (RASSF) have included the presence of heavy metals and adulteration with non-permitted dyes.

The International Agency for Research on Cancer (IARC) classified glycidol as probably carcinogenic to humans (group 2A) and 3-MCPD as a possible human carcinogen (group 2B).[5] The parent MCPD/glycidol moiety may be released from the fatty acid esters of these compounds following ingestion as a result of digestive processes making the compounds available for uptake from the gut. The esters are metabolised to free 3-MCPD and free glycidol in rodents and present adverse toxicological effects and a similar behaviour and impact may be anticipated in humans.[6]

Commission Regulation EC No. 1881/2006 addresses 3-MCPD, 2-MCPD and GE where EFSA have set the tolerable daily intake (TDI) for 3-MCPD and its fatty acid esters (expressed as MCPD equivalents) to 2 μg kg^{-1} bodyweight per day. The recent EFSA Scientific Opinion (2016), stated that high consumers of at-risk foods had a 3-MCPD intake that exceeded the TDI fourfold.[7] In the case of infants consuming formula milk, the estimated exposure to 3-MCPD was 2.4 μg kg^{-1} bodyweight per day using mean occurrence and 3.2 μg kg^{-1} bodyweight per day using the 95th percentile of occurrence. It was recommended that the TDI was reduced further to 0.8 μg kg^{-1} bodyweight per day.[7]

1.4 MCPD

MCPD formation occurs in acid-hydrolysed vegetable protein (HVP) and formation is due to a reaction between hydrochloric acid and lipids that occurs more rapidly at the high temperatures used in processing.

In thermally processed foods such as bakery products, malt-derived products, cooked/cured fish or meat, formation is by heat as a reaction product of triacylglycerols, phospholipids or glycerol and hydrochloric acid (usually from sodium chloride naturally present or added) in fat-based or fat-containing foods, especially with oil temperatures of 200 °C or above.

For 3-MCPD and 2-MCPD the at-risk categories include:

- Refined vegetable oils (especially palm oil), margarines
- Infant formula, follow-on formula and foods for special medical purposes intended for infants and young children
- Malt-derived products
- Cooked/cured fish or meat
- Bakery goods including pastries and cakes

EFSA released an update (EFSA, 2018) of the risk assessment on 3-MCPD in vegetable oils and food to reassess the possible long-term adverse effects on the kidney and male fertility.[8] It was noted that the TDI of 2 $\mu g\ kg^{-1}$ per day established by EFSA in 2017 is not exceeded in the adult population, though a slight exceedance of the TDI was observed in high consumers of younger age groups, particularly infants receiving formula only.

1.5 MCPD Esters

During the production of fats and oils, fatty acid esters may be formed from 2-MCPD, 3-MCPD and glycidol when the fats and oils are heated to high temperatures in the presence of chloride ions.

For MCPD esters the at-risk foods are:

- Refined vegetable oils
- Infant formula, follow-on formula and foods for special medical purposes intended for infants and young children

The concentrations of 3-MCPD esters are lowest in rapeseed oil and highest in palm oil, a widely used food ingredient. MCPD esters are formed from acylglycerols in the presence of chlorinated compounds during deodorization at high temperature (generally >200 °C).[9] Both natural and synthetic anti-oxidants present during deep fat frying of potato chips in refined, bleached and deodorized palm oil have been shown to reduce the occurrence of 3-MCPD and glycidyl esters, by inhibiting the formation of radical intermediates.[10]

1.6 Glycidyl Esters (GEs)

GEs are formed from diacylglycerols with no requirement for chlorinated compounds. GE formation is directly associated with elevated temperatures (>240 °C) and the time at these elevated temperatures. In particular, refined vegetable oil deodorization at elevated temperatures >240 °C is the critical step in the formation of GE.

For glycidyl esters, the at-risk foods listed in Commission Regulation (EC) No 1881/2006 are:

- Vegetable fats and oils, margarines
- Infant formula, follow-on formula and foods for special medical purposes intended for infants and young children
- Products containing refined vegetable oils, especially refined palm oil

Commission Regulation (EU) 2018/290 was published in February 2018 and came into force in March 2018. This provided for an amendment to Regulation (EC) No 1881/2006 concerning the maximum levels of glycidyl fatty acid esters permitted in certain foods. Commission Regulation (EU) 2018/290 covers edible oils and infant formula as well as soy sauce and hydrolysed vegetable protein. Limits for glycidyl esters were set for infant formula (powder and liquid) at 5 $\mu g\ kg^{-1}$ and 10 $\mu g\ kg^{-1}$ respectively.

1.7 Furan and Methylfurans

The IARC has classified furan as a possible human carcinogen (Group 2B).[5] Furan and related compounds 2- and 3-methyl furan are formed in foods during thermal processing and can co-occur. Furans are formed unintentionally in heat processed foods as products of the Maillard reaction, particularly in coffee and in canned or jarred foods. Multiple precursors have been identified that are natural food components, including carbohydrates, amino acids, polyunsaturated fatty acids and carotenoids.[11,12]

Commission Regulation (EC) No. 196/2007 addresses furan and related compounds in food which includes 2-methylfurans and 3-methylfurans. For furan, 2-methylfurans and 3-methylfurans the at-risk categories are thermally processed items which primarily include:

- Coffee – solid, instant, roasted beans, ground roasted beans
- Canned and jarred foods including baby food containing meat and various vegetables

The final concentrations of furan depend on multiple factors such as industrial or home processing, preparation for consumption (type of brewing, microwave or stove heating) and individual consumption habits. The EFSA Scientific Opinion of October 2017 concluded that the level of exposure to furans in food indicates a potential human health concern.[13] Data from animal exposure studies that was taken into consideration indicated that liver damage and liver cancer were the most significant health effects. The EFSA noted that the Joint Food and Agriculture Organisation of the United Nations (FAO) and World Health Organisation (WHO) Expert Committee on Food Additives (JECFA) had also concluded that the level of exposure to furan in food indicates a human health concern. Based on these observations and conclusions, furans are an important public health concern and therefore, similar to acrylamide, food processing by FBOs and consumer behaviour will both have a key role in the subsequent formation of furans and likelihood of dietary exposure.

1.8 Polycyclic Aromatic Hydrocarbons

Polycyclic aromatic hydrocarbons (PAHs) are a group of persistent organic pollutant compounds that are composed of two or more aromatic rings, generally produced as a result of incomplete combustion or pyrolysis of organic materials. More than one hundred different PAH compounds exist and these ae found primarily as mixtures. PAHs enter the environment as a result of anthropogenic activities, for example vehicle exhausts emissions or incineration. The major route of human exposure to PAHs for non-smokers is from food.[14,15]

Food can be contaminated with PAHs derived from environmental sources, or by industrial food processing methods, for example heating, drying and smoking and during domestic food preparation, for example smoking, drying, grilling, barbecuing, toasting and roasting. For PAHs, the at-risk categories are thermally processed items and where exposure to the products of fossil fuel or organic material combustion take place such as smoking or drying.[15,16] These include:

- Oils and fats
- Cocoa beans and derived products *e.g.* cocoa fibre
- Smoked/dried meat, fish and shellfish and derived products
- Dried herbs and spices
- Banana chips
- Processed cereal-based foods and baby foods
- Infant formulae and follow-on formulae and milks
- Food supplements

The EFSA Scientific Opinion (2008) reported the findings of their evaluation of the potential toxic threat of dietary exposure to PAHs.[17] Some PAHs such as benzo[*a*]pyrene and benzo[*b*]fluoranthene were recognised as genotoxic carcinogens and consequently dietary exposure should be kept as low as reasonably practicable. Four marker PAH compounds (benzo[*a*]pyrene, chrysene, benz[*a*]anthracene and benzo[*b*]fluoranthene) were identified which became the focus of the subsequent regulation of which Commission Regulation (EC) No. 1881/2006 was first. Further regulations that have addressed PAHs in food have included Commission Regulation (EU) No. 420/2011 (which prescribed the maximum sum of the four marker PAHs; PAH4), Commission Regulation (EU) 2015/1125 and Commission Regulation (EU) 2015/1933.

The effects of time, cooking method, pH and marinades on PAH formation have been reported.[18,19] Vegetable oils have been shown to contain PAHs and extended frying times, or repeated use of the oils, have resulted in elevated concentrations.[20,21] Consequently, the repeated use of cooking oils should be avoided and this further suggests a requirement for suitable guidance to FBOs and actions to inform and modify consumer behaviour to minimise exposure.

1.9 Ethyl Carbamate

Ethyl carbamate (EC), also known as urethane, is an ethyl ester of carbamic acid. EC can be formed from various substances derived from food and beverages, including hydrogen cyanide, urea, citrulline, and other *N*-carbamyl compounds. Cyanate is probably the ultimate precursor in most cases, reacting with ethanol to form the carbamate ester.[22]

EC has been classified as a probable carcinogen (group 2A) by the IARC.[5] EC has been linked to lung cancer and mutations in organs such as liver, mammary glands, heart, ovary and stomach. EC exposure has been linked to neurological disorders and exposure during pregnancy can cause damage to the developing foetal nervous system.[23]

EC has been found in many fermented food products and alcoholic beverages although stone fruit brandies present the highest risk of ethyl carbamate formation. For some wines, production and storage conditions may also elevate the likelihood of ethyl carbamate formation, *e.g.* the use of urea-based fertilizers, exposure of the fermented product to light and presence of fruit must or pomace in the bottles during storage.

Commission Regulation (EC) No.110/2008 addressed ethyl carbamate in food and beverages and at-risk categories identified included:

- Spirits, wine and beer, especially stone fruit brandies
- Fermented foods and beverages, including soy sauce
- Yoghurt

On the basis of animal studies, JECFA concluded that EC is a genotoxic carcinogen.[24] The EFSA has since requested data on levels of EC compounds in foods and beverages and the data gathered will be used to undertake a risk assessment of the threat to population health. Further regulation to control the levels of EC in at risk products and ensure consumer protection may therefore be anticipated.

1.10 Heterocyclic Amines

Heterocyclic amines (HCAs) are formed when amino acids, sugars and creatine or creatinine react at temperatures in excess of 150 °C. HCAs are formed by cooking protein-rich foods, especially those containing animal muscle tissues such as meat and fish, at high temperatures such as frying, grilling or barbecuing directly over an open flame. The levels of HCAs produced in meat products are mainly dependent on the kinds of meat product, cooking time and temperature. These HCAs are not only present in cooked red meat, fish, and chicken, but are also present at lower levels in baked and fried foods derived from grain.[25]

The IARC has categorized the following heterocyclic amines as possible human carcinogens:[5] 2-amino-1-methyl-6-phenylimidazo[4,5-*b*]pyridine (PhIP), 2-amino-3,8-dimethylimidazo[4,5-*f*]quinoxaline (MeIQx), and 2-amino-3,4-dimethylimidazo[4,5-*f*]quinoline (MeIQ), whereas 2-amino-3-methylimidazo[4,5-*f*]quinoline (IQ) has been categorized as a probable human carcinogen. Although not subject to specific regulation at present, HCAs have been shown to form simultaneously with PAHs in grilled meats.[26] Consequently, existing EU regulations to control consumer exposure to PAHs will prove beneficial.

1.11 5-Hydroxymethylfurfuryl

5-Hydroxymethylfurfural (HMF) is a heterocyclic aldehyde, formed from sugars such as glucose or fructose as an intermediate of the Maillard reactions that take place during thermal processing, or subsequently during storage. HMF is found in many thermally processed foodstuffs, for example bakery products, cookies, candies, chewing gums and alcoholic beverages. HMF may be present due to the use of corn syrups and may also be present in fruit juices, dried fruits, crackers, vinegars, baby food, honey and coffee. In carbonated soft drinks, the sources of HMF are both sweeteners and sulphite ammonia caramel, added as a food colouring. The levels of this compound depend on the type of drink and sweetener used, for example sucrose or glucose–fructose syrup.[27]

HMF is a quality marker in honey and concentrations increase due to long-term storage under inappropriate conditions, excessive heating to achieve decrystallisation, or as a result of adulteration with invert sugars. Additional factors include the honey variety, chemical composition and acidity.[27,28] Council Directive 2001/10/EC and subsequent amendments address the organoleptic and chemical properties of honey permitted for sale in the EU. At present, a maximum permitted level for 5-hydroxymethylfurfural of 40 mg kg^{-1} in honey has been established.

HMF exhibits cytotoxicity towards the mucous membranes, skin and upper respiratory tract, as well as mutagenicity and carcinogenicity with observable chromosome abnormalities. In contrast it has been reported to exhibit beneficial anti-oxidant, anti-inflammatory, anti-hypoxic and anti-cell sickling properties and have therapeutic uses.[28–30] At present it has not been fully confirmed that dietary HMF presents a risk to consumer health.

1.12 Additional Food Processing-based Risks

Fermentation (alcoholic, acidic, alkaline), pickling, salting and putrefaction present a diverse range of potential additional chemical safety issues. Risks associated with fermented foods can arise from incorrect or contaminated starter cultures, incorrect or poorly regulated fermentation conditions and contaminated fermentation substrates. Contaminants which may be present include biogenic amines, *N*-nitroso compounds (NOCs), for example nitrosamine and mycotoxins (for example aflatoxin and ochratoxin).[31] At risk foods include:

- Raw fermented sausage *e.g.* salami, pepperoni, spreads, luncheon meats
- Pickled vegetables
- Sauerkraut
- Shrimp paste/shrimp sauce
- Fish sauce (Yulu)
- Chocolate
- Soy sauce
- Bean curd (fermented tofu – coagulated soy milk)
- Douchi (fermented and salted black soybean)
- Miso (fermented soy beans)
- Natto (fermented soy beans)
- Tempeh (fermented soy beans)
- Kimchi (fermented cabbage)
- Kombucha (fermented tea)
- Idli (fermented black lentils)
- Ogiri (fermented melon)
- Iru (fermented locust bean)
- Ugba (fermented African oil bean)
- Mahewu (fermented maize meal)
- Ogi (fermented maize meal)

In addition to fermentation, incorrect handling and storage of produce may also result in contamination with the metabolic products from putrefaction (decay) processes. Contaminant substances examples include ethylamine, putrescine, cadaverine, histamine, tyramine, tryptamine, phenylethylamine. At risk sources include:

- Wine
- Cheese
- Fish
- Meat

- Vegetables, for example sauerkraut
- Other traditional foods often not widely consumed and with restricted geographic distribution, for example Surstromming (fermented herring)

1.13 Conclusions

The processing conditions most likely to generate chemical safety risks for foods include:

- Thermal processing >120 °C for high asparagine, low moisture materials, for example baking, roasting, frying and drying
- Exposure to the combustion products of fossil fuels and organic materials *e.g.* smoking and drying
- Heating of oils or fats at elevated temperatures >200 °C, especially in the presence of chlorinated components
- Fermentation (alcoholic, acidic, alkaline), pickling, salting and putrefaction

Whilst many of the priority process contaminants are subject to specific regulation within the EU, there are several which are not, despite their known or suspected risk to consumer health. Although not subject to specific regulation to control or minimise exposure risk, some are currently subject to investigation by EFSA. For example, advanced glycation end products (AGEs) have been highlighted as a prominent area for possible future legislative measures, as currently there is no AGE-related legislation, although EFSA do have AGE levels in food on their watchlist. EFSA recently identified future research priorities under the theme: Protecting consumers' health with independent scientific advice on the food chain. The full EFSA Report can be found at this link: (EFSA Strategy 2020 – Trusted science for safe food): https://www.efsa.europa.eu/sites/default/files/corporate_publications/files/strategy2020.pdf.

EU Regulations

General

Regulation (EC) No 178/2002 of the European Parliament and of the Council of 28 January 2002; laying down the general principles and requirements of food law, establishing the European Food Safety Authority and laying down procedures in matters of food safety.

Regulation (EC) No 882/2004 of the European Parliament and of the Council of 29 April 2004; on official controls performed to ensure the verification of compliance with feed and food law, animal health and animal welfare rules.

Contaminants

Council Regulation (EEC) No 315/93 of 8 February 1993; laying down Community procedures for contaminants in food.

Commission Regulation (EC) No 466/2001 of 8 March 2001 setting maximum levels for certain contaminants in foodstuffs.

Commission Regulation (EC) No 2065/2003 of the European Parliament and of the Council of 10 November 2003; on smoke flavourings used or intended for use in or on foods.

Commission Regulation (EC) No 627/2006 of 21 April 2006 implementing Regulation (EC) No 2065/2003 of the European Parliament and of the Council; as regards quality criteria for validated analytical methods for sampling, identification and characterisation of primary smoke products.

Commission Regulation (EC) No 1881/2006 of 19 December 2006; setting maximum levels for certain contaminants in foodstuffs.

Commission Regulation (EC) No 196/2007 of 26 February 2007 fixing the export refunds on cereals and on wheat or rye flour, groats and meal.

Commission Regulation (EC) No 333/2007 of 28 March 2007; laying down the methods of sampling and analysis for the official control of the levels of lead, cadmium, mercury, inorganic tin, 3-MCPD and benzo(*a*)pyrene in foodstuffs.

Commission Regulation (EC) No 110/2008 of the European Parliament and of the Council of 15 January 2008 on the definition, description, presentation, labelling and the protection of geographical indications of spirit drinks and repealing Council Regulation (EEC) No 1576/89

Commission Regulation (EC) No 629/2008 of 2 July 2008; amending Regulation (EC) No 1881/2006 setting maximum levels for certain contaminants in foodstuffs.

Commission Regulation (EU) No 420/2011 of 29 April 2011; amending Regulation (EC) No 1881/2006 setting maximum levels for certain contaminants in foodstuffs.

Commission Regulation (EU) No 835/2011 of 19 August 2011; amending Regulation (EC) No 1881/2006 as regards maximum levels for polycyclic aromatic hydrocarbons in foodstuffs.

Commission Regulation (EU) No 836/2011 of 19 August 2011 amending Regulation (EC) 333/2007; laying down the methods of sampling and analysis for the official control of levels of lead, cadmium, mercury, inorganic tin, 3-MCPD and benzo(*a*)pyrene in foodstuffs.

Commission Regulation (EU) No 1169/2011 of the European Parliament and of the COUNCIL of 25 October 2011; on the provision of food information to consumers, amending Regulations (EC) No 1924/2006 and (EC) No 1925/2006 of the European Parliament and of the Council, and repealing Commission Directive 87/250/EEC, Council Directive 90/496/EEC, Commission Directive 1999/10/EC, Directive 2000/13/EC of the European Parliament and of the Council, Commission Directives 2002/67/EC and 2008/5/EC and Commission Regulation (EC) No 608/2004.

Commission Regulation (EU) No 1258/2011 of 2 December 2011 amending Regulation (EC) No 1881/2006; setting maximum levels for certain contaminants in foodstuffs.

Commission Regulation (EU) 2015/1125 of 10 July 2015 amending Regulation (EC) No 1881/2006 as regards maximum levels for polycyclic aromatic hydrocarbons in Katsuobushi (dried bonito) and certain smoked Baltic herring.

Commission Regulation (EU) 2015/1933 of 27 October 2015 amending Regulation (EC) No 1881/2006 as regards maximum levels for polycyclic aromatic hydrocarbons in cocoa fibre, banana chips, food supplements, dried herbs and dried spices.

Commission Regulation (EU) 2017/2158 of 20 November 2017; establishing mitigation measures and benchmark levels for the reduction of the presence of acrylamide in food.

Commission Regulation (EU) 2018/290 of 26 February 2018 amending Regulation (EC) No 1881/2006; as regards maximum levels of glycidyl fatty acid esters in vegetable oils and fats, infant formula, follow-on formula and foods for special medical purposes intended for infants and young children.

References

1. N. G. Halford and T. Y. Curtis, *Acrylamide in Food*, World Scientific, London, 2019, p. 172, ISBN: 978-1-78634-659-9.
2. M. Mesias, C. Delgado-Andrade, F. Holgado and F. J. Morales, Acrylamide content in French fries prepared in households: A pilot study in Spanish homes, *Food Chem.*, 2018, **260**, 44–52.
3. EFSA, EFSA CONTAM Panel (EFSA Panel on Contaminants in the Food Chain) 2015, Scientific Opinion on acrylamide in food, *EFSA J.*, 2015, **13**(6), 4104, 321 pp.
4. B. Q. Lee and S. M. Khor, 3-Chloropropane-1,2-diol (3-MCPD) in Soy Sauce: A Review on the Formation, Reduction, and Detection of This Potential Carcinogen, *Compr. Rev. Food Sci. Food Saf.*, 2015, **14**(1), 48–66.
5. IARC, *Agents Classified by the IARC Monographs, Volumes 1–109*, 2014, Available from, http://monographs.iarc.fr/ENG/Classification/Classificat-ions GroupOrder.pdf.
6. J. H. Spungen, S. MacMahon, J. Leigh, B. Flannery, G. Kim, S. Chirtel and D. Smegal, Estimated US infant exposures to 3-MCPD esters and glycidyl esters from consumption of infant formula, *Food Addit. Contam., Part A*, 2018, **35**(6), 1085–1092.
7. EFSA, EFSA CONTAM Panel (EFSA Panel on Contaminants in the Food Chain), 2016. Risks for human health related to the presence of 3- and 2-monochloropropanediol (MCPD) and their fatty acid esters and glycidyl fatty acid esters in food, *EFSA J.*, 2016, **14**(5), 4426, 159 pp.

8. EFSA, EFSA CONTAM Panel (EFSA Panel on Contaminants in the Food Chain), 2018b. Update of the risk assessment on 3-monochloropropane diol and its fatty acid esters, *EFSA J.*, 2018, **16**(1), 5083, 48 pp.
9. J. Smidrkal, M. Tesarova, I. Hradkova, M. Bercikova, A. Adamcikova and V. Filip, Mechanism of formation of 3-chloropropan-1,2-diol (3-MCPD) esters under conditions of the vegetable oil refining, *Food Chem.*, 2016, **211**, 124–129.
10. Y. H. Wong, K. M. Goha, K. L. Nyam, I. A. Nehdi, H. M. Sbihi and C. P. Tan, Effects of natural and synthetic antioxidants on changes in 3-MCPD esters and glycidyl ester in palm olein during deep-fat frying, *Food Control*, 2019, **96**, 488–493.
11. M. S. Altaki, F. J. Santos, L. Puignou and M. T. Galceran, Furan in commercial baby foods from the Spanish market: estimation of daily intake and risk assessment, *Food Addit. Contam., Part A*, 2017, **34**(5), 728–739.
12. C. Rannou, D. Laroque, E. Renault, C. Prost and T. Serot, Mitigation strategies of acrylamide, furans, heterocyclic amines and browning during the Maillard reaction in foods, *Food Res. Int.*, 2016, **90**, 154–176.
13. EFSA, EFSA CONTAM Panel (EFSA Panel on Contaminants in the Food Chain) Scientific opinion on the risks for public health related to the presence of furan and methylfurans in food, *EFSA J.*, 2017, **15**(10), 5005, 142 pp.
14. H. I. Abdel-Shafy and M. S. M. Mansour, A review on polycyclic aromatic hydrocarbons: Source, environmental impact, effect on human health and remediation, *Egypt. J. Pet.*, 2016, **25**(1), 107–123.
15. J. L. Domingo and M. Nadal, Human dietary exposure to polycyclic aromatic hydrocarbons: A review of the scientific literature, *Food Chem. Toxicol.*, 2015, **86**, 144–153.
16. J. Li, H. Dong, X. Li, B. Han, C. Zhu and D. Zhang, Quantitatively assessing the health risk of exposure to PAHs from intake of smoked meats, *Ecotoxicol. Environ. Saf.*, 2016, **124**, 91–95.
17. EFSA, EFSA CONTAM Panel (EFSA Panel on Contaminants in the Food hain), Scientific Opinion of the Panel on Contaminants in the Food Chain on a request from the European Commission on Polycyclic Aromatic Hydrocarbons in Food, *EFSA J.*, 2008, **724**, 1–114.
18. K. C. Park, H. Pyo, W. Kim and K. S. Yoon, Effects of cooking methods and tea marinades on the formation of benzo(*a*)pyrene in grilled pork belly (Samgyeopsal), *Meat Sci.*, 2017, **129**, 1–8.
19. W. Wongmaneepratip and K. Vangnai, Effects of oil types and pH on carcinogenic polycyclic aromatic hydrocarbons (PAHs) in grilled chicken, *Food Control*, 2017, **79**, 119–125.
20. X. W. Hao, J. Li and Z. L. Yao, Changes in PAHs levels in edible oils during deep-frying process, *Food Control*, 2016, **66**, 233–240.
21. Z. Zelinkova and T. Wenzl, The occurrence of 16 EPA PAHs in food- A review, *Polycyclic Aromat. Compd.*, 2015, **35**(2–4), 248–284.
22. Y. G. Kim, J. Lyu, M. K. Kim and K.-G. Lee, Effect of citrulline, urea, ethanol, and urease on the formation of ethyl carbamate in soybean paste model system, *Food Chem.*, 2015, **189**, 74–79.

23. V. Gowd, H. M. Su, P. Karlovsky and W. Chen, Ethyl carbamate: An emerging food and environmental toxicant, *Food Chem.*, 2018, **248**, 312–321.
24. JECFA (2006) Evaluation of certain food contaminants, 64th report of the Joint FAO/WHO Expert Committee on Food Additives, World Health Organisation, Geneva, ISBN 92 4 120930 5.
25. S. A. O. Adeyeye, Heterocyclic amines and polycyclic aromatic hydrocarbons in cooked meat products: A review, *Polycyclic Aromat. Compd.*, 2018, 11, DOI: 10.1080/10406638.2018.1559208.
26. N. H. A. Kamal, J. Selamat and M. Sanny, Simultaneous formation of polycyclic aromatic hydrocarbons (PAHs) and heterocyclic aromatic amines (HCAs) in gas-grilled beef satay at different temperatures, *Food Addit. Contam., Part A*, 2018, **35**(5), 848–869.
27. U. M. Shapla, M. Solayman, N. Alam, M. I. Khalil and S. H. Gan, 5-Hydroxymethylfurfural (HMF) levels in honey and other food products: effects on bees and human health, *Chem. Cent. J.*, 2018, 12–35.
28. A. Obiedzinska, E. Biller, M. W. Obiedzinska and M. Tyszka, Determination of hydroxymethylfurfural content in natural honeys in Poland, *Pol. J. Appl. Sci.*, 2018, (4), 60–64.
29. S. Kowalski, M. Lukasiewicz, A. Duda-Chodak and G. Ziec, 5-hydroxymethyl-2-furfural (HMF) – heat induced formation, occurrence in food and biotransformation: A review, *Pol. J. Food Nutr. Sci.*, 2013, **63**(4), 207–225.
30. F. Kong, H. L. Bae and K. Wei, 5-Hydroxymethylfurfural Mitigates Lipopolysaccharide-Stimulated Inflammation via Suppression of MAPK, NF-κB and mTOR Activation in RAW 264.7 Cells, *Molecules*, 2019, **24**(2), 275.
31. M. P. Costa, B. L. Rodrigues, B. S. Frasao and B. A. Conte-Junior, Biogenic Amines as a Food Quality Index and Chemical Risk for Human Consumption, in *Food Quality: Balancing Health and Disease*, ed. A. M. Holban and A. M. Grumezescu, Handbook of Food Bioengineering, Academic Press, 2018, vol. 13, pp. 75–108.

CHAPTER 2

Formation, Analysis, Occurrence and Mitigation of Acrylamide Content in Foods

P. ŠIMKO* AND L. KOLARIČ

Slovak University of Technology in Bratislava, Institute of Food Science and Nutrition, Faculty of Chemical and Food Technology, Radlinského 9, 812 37, Slovak Republic
*E-mail: qsimko@stuba.sk

2.1 Introduction

There is no doubt that the development of mankind in society has been tightly confined by access to sufficient amounts of food able to satisfy vital and functional needs of the human organism. However, while mankind took care about food quantity in the past, today the situation is different, and interest is paid to food quality and safety attributes. It is clear that this in turn has been stimulated by development in scientific disciplines such as inorganic, organic and physical chemistry, biochemistry, chemical, physical, sensorial and DNA analysis, nutrition, food chemistry and health sciences, and – finally – risk assessment procedures. Today, all these disciplines generate data about food composition including information regarding hazardous compounds occurring in foods. Because the negative impact of hazardous compounds to human health has been shown in numerous studies, there is

Food Chemistry, Function and Analysis No.19
Mitigating Contamination from Food Processing
Edited by Catherine S. Birch and Graham A. Bonwick
© The Royal Society of Chemistry 2020
Published by the Royal Society of Chemistry, www.rsc.org

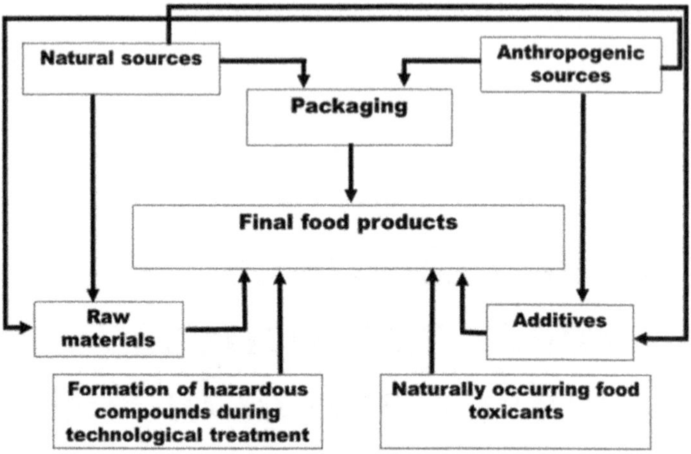

Figure 2.1 Sources and possible routes of food contamination by hazardous compounds.

a strong need to find procedures for the production of safe foods to protect consumers against exposure to these compounds. Therefore, food scientists thoroughly research processes leading to the minimisation of hazardous compounds content in foods to protect the health of humans as much as possible. It is obvious that this search is complicated since food contamination can occur at any time and at any stage of food production, or storage, respectively.

As follows from Figure 2.1, there are, in principle, four sources of hazardous compounds that may contaminate foods. The first is confined to natural sources such as volcanic activity, forest fires, *etc.* The second source is anthropogenic in nature and correlates with industrial activity (crude oil processing and transportation, chemical industry, production of energy, heavy industry, traffic, *etc.*). The third source is associated with the food matter itself, when some toxicants (*e.g.* glycoalkaloids, bicyclic octapeptides, allyl-substituted benzene, *etc.*) are formed in raw materials as protectants against pests or invasive microorganisms, respectively. Finally, the fourth one is associated directly with food treatment (especially at increased temperatures) such as frying, baking, smoking, drying and handling including packaging and storage. These processes generate hazardous compounds such as *N*-nitrosamines, heterocyclic amines, polycyclic aromatic hydrocarbons, biogenic amines, hydroperoxides, benzene, Maillard reaction products (*e.g.* acrylamide, 2-hydroxymethyl-5-furaldehyde, furane), *etc.* Hence, knowing detailed information regarding the formation and behaviour of these compounds in the food matrix is fundamental in preventing their generation, or minimising their content.[1] So, in relation to the above-mentioned aspects, this chapter summarises information and data from experimental studies conducted regarding the formation, analysis, occurrence and mitigation of acrylamide (AA) content in foods.

2.2 Formation of AA in Foods

The first relevant information about the formation of AA in foods was released in 2002 when Swedish researchers published results of their observations relating to a thermally treated food matrix, especially within a matrix rich in saccharide content such as potato, beetroot, and also certain heated commercial potato products and crispbread.[2] As found, AA can be formed during the thermal treatment of certain amino acids (mainly asparagine), particularly in combination with reducing saccharides, resulting in early Maillard reaction products called *N*-glycosides.[3] The key role of asparagine in the production of AA by Maillard reaction was also confirmed by Mottram *et al.* in 2002.[4] The Maillard reaction, also denoted as 'non-enzymatic browning', is a long known thermal reaction responsible for the brown colour of processed food and the formation of a multitude of characteristic flavour compounds formed during food processing of bread, meat, coffee, nuts and other food stuffs at higher temperatures.[5] However, besides the formation of these desirable substances, the Maillard reaction was also identified as the formation mechanism of a number of undesirable compounds, of which AA is just one. However, data on other undesirable heat-processed products are too limited to perform a group evaluation for all the Maillard reaction products that are formed in addition to AA.[6] Also, other findings indicated that the Maillard reaction-driven generation of flavour and colour in thermally processed foods (treated above 120 °C and at low moisture) can be associated, in general, with the formation of AA.[7] On the other hand, some studies indicate formation of substantial amounts of AA in model systems even at temperatures lower than 100 °C under conditions that resemble the drying of foods, such as prunes. It was concluded that AA in prunes and prune juice very likely originates from saccharides and asparagine that are present in considerable amounts in the starting material, *i.e.* plums.[8] AA can also be generated from 3-aminopropionamide by its deamination, a transient intermediate formed during thermal degradation or enzymatic decarboxylation of asparagine.[9] Also, the occurrence of 3-aminopropionamide in various amounts was observed in several potato cultivars.[10] Other pathways of AA formation that do not require asparagine are also described in the literature. It was shown that AA can, in principle, be formed from acrolein and acrylic acid, especially in lipid rich foods. Also, the different pathways *via* acrolein and acrylic acid in the formation of AA were described.[11] Similarly, the pyrolytic formation of AA from purified wheat gluten and gluten-supplemented wheat bread rolls was demonstrated.[12] Formation of AA was also observed in sterilized table olives, in model systems based on the aqueous fraction of olive pulp from untreated and lye-treated green olives in commercial model peptides containing protein-bound aspartic acid, alanine and methionine, respectively, at 200 °C, and at other times in the absence of any carbonyl sources. Casado and co-workers concluded that their results strongly support the role of peptides/proteins as precursors of AA formation in sterilized olives.[13]

In addition to AA, glycidamide is also formed during the heat processing of foods, although only to a minor extent. In potato chips (*i.e.* potato crisps), the amount of glycidamide was 0.5% of the amount of AA, whereas this proportion was only 0.2% in French fries, without showing a clear dependence on heating time.[14] In summary, although several pathways of AA formation in food were investigated and demonstrated, especially in model systems, AA is formed predominantly from free amino acid asparagine and reducing saccharides (glucose and fructose) during high-temperature cooking, such as frying, roasting and baking, and processing.[15]

2.3 Chemistry of AA Formation

2.3.1 Chemical Characteristics of AA

AA (CAS No 79-06-01) is an odourless white crystalline solid with the molecular formula C_3H_5NO and a molecular weight of 71.08 g mol^{-1}. Synonyms are *inter alia* 2-propenamide, acrylic amide and ethylene carboxamide. The structural formula of AA is shown in Scheme 2.1.

The melting point of AA is 84.5 °C and the vapour pressure is 0.9 Pa (7×10^{-3} mm Hg) at 25 °C. Solubility at 30 °C is high in water (2155 g L^{-1}), methanol (1550 g L^{-1}), ethanol (862 g L^{-1}), and acetone (631 g L^{-1}). AA is less soluble in chloroform (26.6 g L^{-1}) and benzene (3.46 g L^{-1}). The log K_{OW} is −0.67, the log k_{oc} is 1, and Henry's law constant at 25 °C is 1.7×10^{-9} atm m^{-3} mol^{-1}.[6] AA is stable at room temperature but readily polymerizes if heated to melting point or if exposed to ultraviolet radiation. The results showed that AA was quite stable in aqueous solutions, but much less stable in dry reaction conditions. Buffer type and pH had a significant influence on the decrease in the amount of free AA. The presence of amino acids with a nucleophilic side chain considerably decreased the level of free AA. The highest reactivity was noted for cysteine, leading to the formation of the mono-addition product cysteine-S-β-propionamide, as well as to the double addition product. Other nucleophiles, such as lysine, arginine, serine and ascorbic acid, were less reactive, but yielded comparable condensation products.[16] AA is an α,β-unsaturated carbonyl compound with electrophilic reactivity. It can therefore react with nucleophilic groups of various molecules.[6]

2.3.2 Reaction Schemes of AA Formation

As mentioned above, formation of AA is associated with thermal treatment of a food matrix rich in reducing saccharides and amino compounds. As the main pathway of AA formation in foods is the Maillard reaction,[3,4] the amino acid asparagine has a crucial role in this reaction.[9,17,18] The first

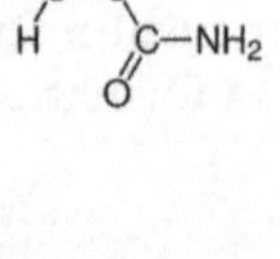

Scheme 2.1

step is the amino–carbonyl interaction between asparagine and a carbonyl group of reducing saccharides at temperatures higher than 100 °C, providing the *N*-glycosylamine of asparagine (also denoted as the initial Schiff base or the *N*-glycoside). While the initial Schiff base is relatively stable under low moisture conditions,[19] in aqueous systems it may hydrolyse to the AA precursors or rearrange to the Amadori compound, which is not an effective precursor in AA formation.[3,4] Most of the initial Schiff base formed rearranges into an Amadori compound, which contributes to colour and flavour formation. The rate limiting step of AA formation is believed to be decarboxylation of the initial Schiff base.[20] Decarboxylation of the initial Schiff base can take place directly to form azomethine ylide, *i.e.* a decarboxylated Schiff base,[9,17] or *via* the intermediate oxazolidine-5-one,[18] as shown in Figure 2.2.

Azomethine ylide rearranges by tautomerization to form Schiff base A and Schiff base B. Schiff base A can be rearranged to form a decarboxylated Amadori compound, or after reaction with a carbonyl compound can provide 3-amino-propionamide or quaternary iminium ion containing compounds, respectively. As can be seen in Figure 2.3, all these compounds are direct precursors of AA.[18]

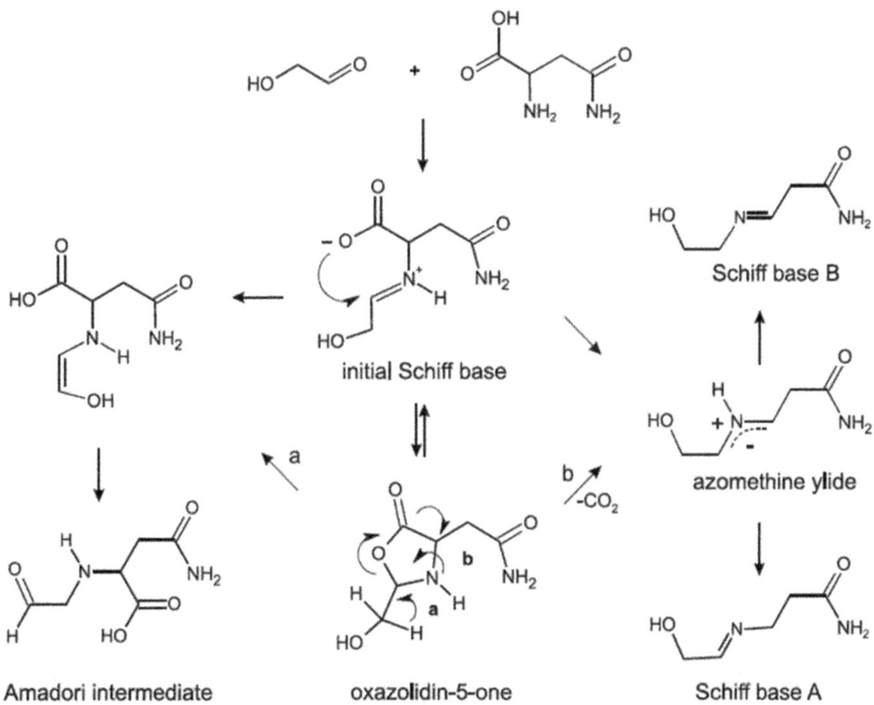

Figure 2.2 The formation of azomethine ylide (for clarity, a simple saccharide glycolaldehyde is used as an example). Reproduced from ref. 18 with permission, Copyright © 2009 VÚP Food Research Institute, Bratislava.

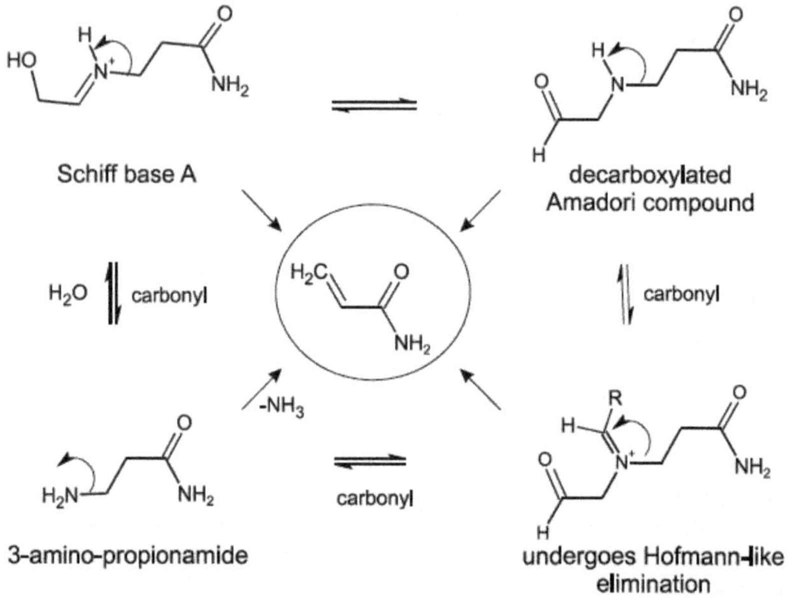

Figure 2.3 Formation of AA from its direct precursors. Reproduced from ref. 18 with permission, Copyright © 2009 VÚP Food Research Institute, Bratislava.

2.3.3 Toxicity of AA

The toxicological properties of AA have been well studied and include neurotoxicity, genotoxicity, carcinogenicity and reproductive toxicity. AA has been classified as a Group 2A carcinogen, *i.e.* probably carcinogenic to humans, by the IARC in 1994.[21] In 2011, the National Toxicology Program determined that AA is reasonably anticipated to be a human carcinogen and the United States Environmental Protection Agency institute characterized AA as 'likely to be carcinogenic to humans'. The American Conference of Governmental Industrial Hygienists has classified AA as an A3 carcinogen, *i.e.* confirmed animal carcinogen with unknown relevance to humans.[22] Experimental results *in vivo* showed that AA is metabolised to a reactive epoxide glycidamide (GA), which is thought to have a major if not exclusive responsibility for the formation of DNA adducts in AA-exposed animals, when GA binding to nucleobases, predominantly adenine and guanine, was observed.[6] Formation of the GA-DNA adduct is shown in Scheme 2.2.

2.4 Legislative Aspects of AA Contained in Foods

In order to protect the health of consumers in EU member states, Article 2 of the Council Regulation (EEC) No 315/93 stipulates that, where necessary, maximum tolerances for specific contaminants shall be established.[23] Thus, a number of maximum tolerances for food contaminants, natural plant toxicants, such as mycotoxins, as well as for process contaminants such as polycyclic aromatic hydrocarbons, 3-monochloropropane-1,2-diol, *etc.* have been

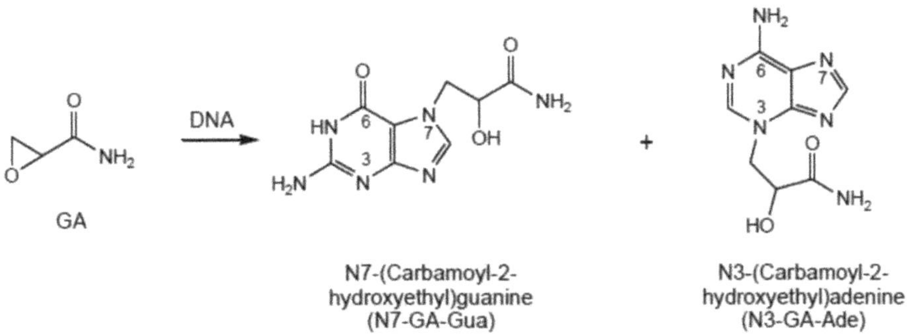

Scheme 2.2

laid down in Commission Regulation (EC) No 1881/2006.[24] Commission Recommendation 2010/307/EU[25] which replaced Commission Recommendation 2007/331/EU[26] recommended that EU member states should perform monitoring of AA levels in certain specified foodstuffs and report the data annually in a prescribed format to the European Food Safety Authority (EFSA). The monitoring studies were targeted especially to those foods that are known to contain high AA content and/or contribute significantly to human dietary intake. Besides the type of product, sampling points and procedure, sample numbers and frequencies, analytical requirements as well as minimum additional information required to be provided for each product, were also each laid down in the two European Commission (EC) recommendations. On the basis of data collected by the EFSA, the European Commission has limited AA content in chosen foods, setting so called 'benchmark levels' (BLs) applicable from 11 April 2018.[27] In contrast with fixed maximum limits of common food contaminants usually used for setting a definitive decision about the safety or hazardousness of foods, BLs are 'only' performance indicators used to verify the effectiveness of the mitigation measures and are based on experience and occurrence of AA content in a broad range of food categories.

2.5 Analysis of AA

Since the initial discovery of AA as a heat-induced food toxicant in fried rat feed in 2000,[28] and then in thermally treated foods in 2002,[2] various methods for the determination of AA content in foods have been developed. The most common methods include chromatography procedures including HPLC tandem mass spectrometry (HPLC-MS/MS) or GC mass spectrometry (GC-MS).[29] The GC step is based either on derivatization (usually bromination to increase relative molecular mass) or on direct analysis without derivatization, respectively. In general, a critical factor for analyte recovery appears to be the extraction of AA from the various food matrices and pre-clean-up of AA fractions.[30] The Association of Official Agricultural Chemists (AOAC) method used for AA analysis is HPLC-MS/MS. However, this method is not economically viable due to the highly expensive MS/MS detector. Forasmuch as MS is believed to be the most suitable technique for GC- and HPLC-based analysis of various food contaminants including AA, development and application of alternative reliable, sensitive, operative, rapid, and low-cost analytical methods able to effectively replace MS/MS techniques would be beneficial especially for food business operators requiring sampling and analysis of AA content in their products on the basis of the requirements set by the EC.[27] Due to the high solubility of AA in water and its high reactivity, in combination with the lack of any chromophore, it is not a particularly easy analytical compound for determination by common GC or HPLC techniques.[31] Furthermore, AA content in foods ranges around μg kg^{-1} levels and therefore sample preparation, including extraction, pre-cleaning and eventual concentration are crucial steps in its analysis.[32]

2.5.1 Sample Preparation

Sample pre-treatment is a usually inevitable step to be applied before final AA determination because of the presence of interfering substances appearing in thermally processed foods. Due to the high diversity of food samples, various sample pre-treatment procedures have been investigated but no universal pre-treatment has been developed so far for all foods. For standard detection methods, such as LC-MS/MS and GC-MS, the general procedures of pre-treatment include homogenization, spiking with an internal standard, extraction, defatting, deproteinization, and purification steps. Figure 2.4 shows the flow chart of the general procedures of sample pre-treatment prior to the detection of AA itself.

Nowadays, the common analytical methods to be applied for quantification of an analyte include external standard and internal standard addition procedures. However, addition of an external standard has shown poor reproducibility in AA analysis.[33] To overcome these problems, samples are spiked with internal standards after homogenization to keep control of the recoveries

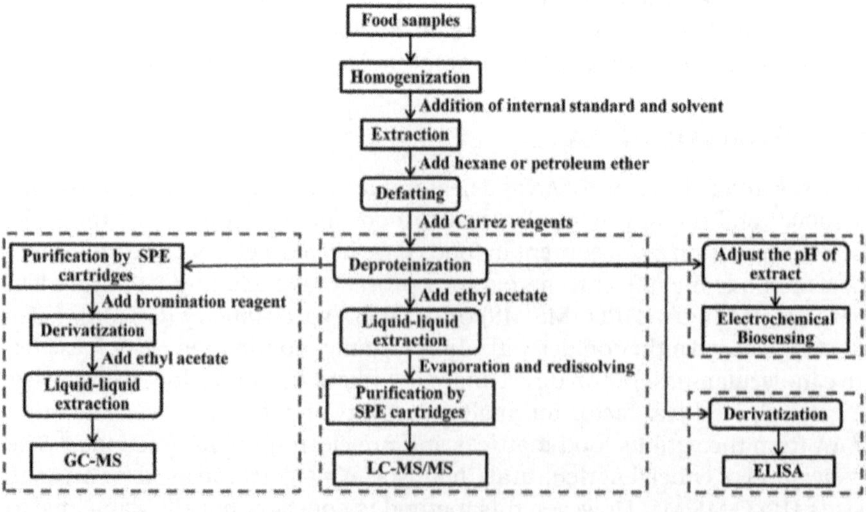

Figure 2.4 Flow chart of general procedures of sample pre-treatment for LC-MS/MS, GC-MS analysis, electrochemical biosensing and ELISA. Reproduced from ref. 33 with permission from Elsevier, Copyright 2015.

achieved and to keep track of possible losses occurring during the whole sample pre-treatment such as extraction and clean-up.[34,35] Usually internal standards for the determination of AA content by MS/MS techniques include isotope-labelled internal standards such as deuterium-labelled $^2H_3$-AA or d_3-AA or carbon-labelled $^{13}C_3$-AA, and non-isotope-labelled internal standards such as methAA, *N*,*N*-dimethylAA, and acetamide, suitable for UV detection.[31,35]

Extraction conditions appear to be the most critical parameter in sample preparation for isolation of AA from a food matrix. Many factors can affect AA extraction yield from food matrices such as sample particle size, defatting, extraction solvent, solvent-to-sample ratio, Ultra Turrax homogenization, the application of mechanical forces, extraction temperature, and time.[35] In the past, various methods were used for the extraction of AA from food samples. These could be divided into two main categories such as extraction with water or with an organic solvent.[36,37] Water at room temperature has been used to extract AA from many kinds of sample matrices in most of analytical methods published so far, as AA has polar characteristics and therefore is soluble in polar solvents including water. Besides water as an extractant, methanol can also be used to extract AA for the convenience of rotatory evaporation and concentration.[36] In numerous studies, other solvents such as ethyl acetate, diethyl ether, acetone and dichloromethane have been examined. However, with respect to the HPLC method to be applied for the determination of AA content, extracted AA was finally transferred to an aqueous phase prior to injection.[37] Some researchers extracted AA selectively by using accelerated solvent extraction (ASE) without co-extracting impurities, which has made possible the simplification of subsequent sample clean-up steps.[38] The results from the extraction optimization procedures indicate that incomplete extraction is the most likely cause of erroneous results in methods applied for the determination of AA content in foods. This might occur when:

- the food matrix is not finely divided enough,
- inappropriate organic solvent is used,
- too short an extraction time and a low extraction temperature is applied,
- all of the above factors occur in combination.

Other possible error sources might be the destruction or formation of AA during the extraction procedure.[29]

The most frequently used purification techniques include the addition of Carrez solutions to separate proteins, extraction with organic solvents to eliminate fats and eventual utilization of solid-phase extraction (SPE), alone or in combination with other purification steps, especially with difficult matrices like coffee, as a clean-up technique of the extracts.[39] SPE has been used extensively in purifying the extracts of food samples due to its simplicity, stability, automation, accuracy, precision, and repeatability of instrumental analysis. Depending on the cartridges used in SPE for purifying the AA fraction, two strategies have been developed. One strategy is based on adsorption of AA from the complex extract in the cartridges through hydrogen bonding, π–π interaction, and cation exchange, followed by elution of AA using other polar solvents. Some common cartridges are Oasis HLB®, Oasis MCX®, IsoluteMulti-Mode®, ENVI-Carb®, and Isolute ENV+®. The second strategy is applied to retain the interferences and collect the elutions containing AA using Oasis HLB® cartridges coupled with Bond Elut-Accucat™ and a homemade SPE column filled with a mixture of C18, SCX, and SAX.[33] In 2006, the QuEChERS (Quick, Easy, Cheap, Effective, Rugged and Safe) method was introduced as a high-throughput extraction method for AA analysis in various food matrices using LC-MS/MS or GC-MS. It involves miniaturized extraction with acetonitrile, liquid–liquid partition by salting out with sodium chloride and magnesium sulphate and clean-up of the extract using a dispersive solid-phase extraction (dSPE) step with loose sorbents such as primary and secondary amines (PSA).[40]

2.5.2 HPLC

HPLC is commonly used in separation and quantification of compounds that are soluble in water and that are either non-volatile under gas chromatography conditions, or are unstable at elevated temperatures, respectively. Determination of the AA content is most often assayed by HPLC coupled with mass spectrometry. In some cases, spectrophotometric detectors with a diode array (UV-DAD) are used.[32] The majority of the proposed HPLC-MS methods make use of a triple quadrupole instrument (MS/MS), and most methods describe electrospray ionization (ESI), although a few have reported good experience with atmospheric pressure chemical ionization (APCI), *e.g.*, for coffee and cocoa, potato- and cereal-based products, and for more general matrices. The advantage of APCI over ESI is that it is considered less prone to ion suppression.[41] The advantage of the HPLC-MS-based methods is that AA can be analysed without prior derivatization, which considerably simplifies and expedites the analysis. Due to the low molecular mass of AA (71 g mol^{-1}) and thus its low-mass fragment ions, confirmation of the analyte can be achieved with a two-stage MS monitoring of more than one characteristic mass transition. However, AA is a very polar molecule with poor retention on conventional HPLC reversed-phase sorbents, and despite the use of tandem MS more effort may need to be placed on efficient clean-up steps to avoid interference from co-extractives.[42] HPLC-UV has been used to determine AA contents in food products containing high levels of AA, where sensitivity is not a major issue. Because of a lack of a chromogenic group, UV measuring has to be performed at low wavelength (195–205 nm), which is also a suitable wavelength for a whole range of other unsaturated molecules. This fact, finally, results in low selectivity during the analysis of AA appearing in processed foods at very low levels.[35] Wang *et al.*,[43] applied a simple and cost-effective method using HPLC coupled with a diode-array detector for determination of AA in baked and deep-fried Chinese foods. The method entails water extraction of AA, sample enrichment and clean-up by solid-phase extraction cartridges followed by detection at 210 nm. The limit of detection and the limit of quantification were 8.0

and 25 μg kg^{-1}, respectively and recoveries of AA ranged between 89.0 and 103%. For chromatographic separation, most researchers use reversed-phase chromatography, *e.g.* Hypercarb column™ (5 μm, Thermo Electron, San Jose, CA, USA) is the most frequently used column in AA analysis. However, it is difficult to choose an appropriate mobile phase to achieve good analyte elution with a reasonable retention time because of the high polarity of AA. Most researchers have experienced problems retaining and separating polar compounds, particularly AA, using conventional reversed-phase chromatography. Waters Atlantis™ columns (Waters, Milford, MA, USA) are designed for these types of challenging separations. Atlantis dC18™ columns are a silica-based line of difunctionally bonded C18 columns that provide the optimal balance of retention for polar and non-polar compounds in reversed-phase chromatography. Atlantis dC18 columns are intelligently designed to provide full LC-MS compatibility, superior peak shapes and excellent column-to-column reproducibility.[36] Examples of pre-separation procedures and HPLC conditions used for the determination of AA in some foods are summarized in Table 2.1.

2.5.3 GC

Quantitative assays of AA contained in foods or its derivatives were also performed by GC-MS with or without a derivatization process.[32,35] Several detection techniques can be coupled with GC for AA determination, such as mass spectrometry (MS), high-resolution MS (HR-MS), ion-trap MS (IT-MS), electron capture detection (ECD), high-resolution time-of flight MS (HR-TOF-MS), micro-electron capture detection (MECD), tandem MS (MS/MS), nitrogen phosphorus detection (NPD), and flame ionization detection (FID). Among these GC techniques the most frequently used in AA analysis in foods is GC-MS.[35] Bromination of the aqueous extract is usually performed in a bromine–water solution with HBr and KBr, or the safer being a KBr + $KBrO_3$ mixture. After derivatization, the excess bromine is neutralized by addition of a sodium thiosulfate solution, and the non-polar derivative is extracted by liquid–liquid extraction into a non-polar organic solvent, *e.g.*, ethylacetate.[41] The bromination of AA has several advantages:

1. in water extracts, bromination produces fewer polar molecules than the native AA, which are easily extracted by ethyl acetate or hexane, eliminating many water-soluble components;
2. bromination enlarges AA molecular weight, which leads to improved MS characteristics, giving several ions to monitor and confirm;
3. the relatively more volatile derivatization product improves GC characteristics.[35]

Some researchers reported the determination of AA using GC-based chromatographic methods without derivatization. Owing to the higher polarity

Table 2.1 Pre-separation procedures and HPLC conditions to be used for determination of AA content in some foods.

Sample	Sample treatment	Column/stationary phase	Mobile phase	Detection	Reference
Biscuits, breakfast cereals	Extraction with water, protein precipitation with Carrez solutions, centrifugation, filtration and pre-separation on Oasis HLB SPE cartridges	Atlantis dC18, 250 × 4.6 mm/5 with pre-column and syringe filters (0.45 μm PVDF)	Water/acetonitrile 96 : 4 (v/v)	UV-DAD	57
Dried fruits, edible seeds	Extraction with water and acetonitrile, addition of prepared QuEChERS mixture, centrifugation, filtration	Gemini RP C18 column (25 cm × 2 mm i.d. × 5 μm particle size × 110 Å pore size)	0.1% Formic acid in water (99.5%, solvent A) and 0.1% formic acid in methanol (0.5%, solvent B)	ESI-MS triple Quadrupole	61
Potato crisps	Extraction with water, centrifugation, filtration	100 × 3.0 mm Hypercarb column with a 10 × 3.0 mm Hypercarb pre-column (both 5 μm particle size)	0.1% Aqueous formic acid	ESI-MS/MS triple quadrupole	62
Potato chips	Extraction with acetone and water, filtration of acetone, evaporation under vacuum, dissolution of the residue	C18-AQ, 2 × 250 mm	Water	UV	37
Wheat-rye, breads, wheat rolls	Extraction with *n*-hexane and water, 2-naphthalenthiol derivatization, centrifugation, filtration	ACE C18 (150 mm × 3 mm; 3 μm)	Acetonitrile and water (60 : 40, v/v)	UV	58
Tea	Extraction with water, addition of acetonitrile and QuEChERS mixture, centrifugation, filtration and preseparation on Oasis MCX SPE cartridge	Reversed ODS-C18 column (250 × 4.6 mm, 5 μm, Hypersil®)	10% Acetonitrile and 90% water containing 0.1% formic acid	MS/MS	60
Cereal-based baby foods	Extraction with 80% methanol in water, defatting with hexane, freezing at −18 °C and cleaning by SPE with Oasis HLB (6 mL, 200 mg) cartridges	Synergi Hydro-RP 80 A 4 μm 250 × 4.6 mm	Isocratic mixture of 5 mM sodium 1-heptanesulfonate in water/acetonitrile (97/3, v/v)	UV-DAD	63
Grilled meat and chicken	Addition of Carrez solution and extraction with 0.2 mM acetic acid solutions, centrifugation, filtration	Zorbax SB-C 18 (4.6 mm × 250 mm, 5 μm)	0.01 mM Acetic acid in 0.2% formic acid and 0.2% acetic acid in acetonitrile (98 : 2 v/v)	MS	64

of the non-derivatized AA, columns with polar phases, *e.g.* polyethylenglycol, have to be used. Temperature programs do not, usually, differ much from those used in methods applying derivatization.[36] A recent GC method employing a HR-TOF-MS analysis was developed for the direct analysis (*i.e.* without derivatization) of AA in various heat-processed foodstuffs.[44] The direct GC-MS analysis of free AA isolated from complex food matrices may be accompanied by several problems, such as:

1. lack of characteristic ions in the mass spectrum of underivatized AA (due to the low molecular weight) and interference caused by the matrix, which increases background noise, and hence a low limit of detection is impossible to obtain;
2. potential generation of AA *in situ* in the heated GC injector from co-extracted precursors (asparagine and reducing sugars) when using water as the extractant solvent;
3. the high solubility of AA in water in comparison with organic solvents complicates sample preparation for GC;
4. owing to the higher polarity of underivatized AA, it has poor retention on conventional GC capillary columns;
5. the compound 3-hydroxypropionitrile may be co-eluted with AA, causing falsely high AA levels.[35]

The most frequently used detectors are mass spectrometers with high resolving power which can record signals generated in selected-ion monitoring (SIM) mode. Another technique which was found to be useful in the determination of AA content is direct solid-phase microextraction (DI-SPME) conducted without derivatization. In this method AA is quantified by GC-PCI-MS/MS in a system of monitoring of selected reactions (SRM). Other detectors used in tandem with gas chromatographs to quantify AA are: flame-ion detectors (FID) or electron capture detectors (ECD). A method using GC-ECD with prior derivatization by $KBrO_3$ and KBr was used to determine amounts of brominated AA derivative in fried foods. ECDs are selective and very sensitive, especially for halogenated molecules.[32] Examples of pre-separation procedures and GC conditions used for the determination of AA in some foods are summarized in Table 2.2.

2.5.4 Rapid Detection Methods

Unlike LC-MS/MS and GC-MS, rapid detection methods are mainly based on the biochemical properties of AA, biomaterials with high and specific interaction with AA, or changes in the physiochemical properties of thermally processed foods related to AA.[33] The colour of the food surface is the first quality parameter evaluated by consumers and it is therefore critical in overall product acceptance. At the same time, colour can be easily assessed as an indicator of other Maillard reaction products.[45] Computer vision software has been used to measure objectively the colour of fried

Table 2.2 Pre-separation procedures and GC conditions to be used for determination of AA content in some foods.

Sample	Sample treatment	Column/stationary phase	Mobile phase	Detection	Reference
Fried potato chips	Homogenization with water, centrifugation, filtration, saturation with NaCl and addition of ethyl acetate, separation of ethyl acetate layer, evaporation, addition of internal standard	DB-WAX fused silica capillary column (0.32 mm i.d. × 30 m length, 0.25 μm film thickness)		GC–nitrogen phosphorus detector (GC–NPD)	31
French fries	Extraction in a water bath 80 °C, filtration, addition of internal standard, treatment with a brominating solution, extraction with ethyl acetate	Capillary column DB17 (50% phenyl; 50% methylsilicone), Agilent (length 30 m × i.d. 0.25 mm)		MS	55
Extruded potato snacks	Extraction with water, addition of internal standard, homogenization, centrifugation, bromination, neutralization of excess bromine, extraction with ethyl acetate, evaporation, clean up on SPE-silica cartridge	Fused-silica column (type DB 5; 25 m × 0.25 mm × 0.25 μm)	Helium	MS/MS	45
Coffee and coffee substitutes	Addition of internal standard, extraction with hot water, addition of acetonitrile and QuEChERS mixture, centrifugation, evaporation, dissolution of residue with acetonitrile, silylation of AA, liquid–liquid extraction	DB-5MS column (30 m × 0.25 mm × 0.25 μm; Agilent Technologies)	Helium	GC/ion-trap MS in single-ion monitoring mode (GC–SIM–MS)	65
Heat-processed starchy, foods (fried instant noodles, non-fried instant noodles, potato chips, and biscuits)	Homogenization, addition of acrylamide working solution and water, filtration, addition of *n*-hexane, bromination, extraction with ethyl acetate, filtration	Suplecowax-10 capillary system (polyethyleneglycol, 30 m × 0.25 mm i.d., 0.25 μm film thickness)	Nitrogen	GC-ECD	34

potatoes since it provides some obvious advantages over a conventional colorimeter, namely, the possibility of analysing the whole surface of the chip, and quantifying characteristics such as brown spots and other defects. The colour of potato chips has been measured using computerized video image processing by means of grey scale values.[46] Detection of AA based on browning is easy to operate and independent from complex instruments. Through taking pictures of food samples and data analysis during processing, the content of AA can be monitored without any pre-treatment. The browning based method is susceptible to many factors, such as intensity of light, shape and flatness of food samples, focal length, aperture, and other parameters of taking photos. Therefore, compared with other novel methods, this method may be more suitable to on-line and real-time preliminary screening of AA during food processing.[33] AA quantification can also be conducted by bioanalytical methods. One of them is enzyme-linked immunosorbent assay (ELISA) in which the identification and quantitative determination is based on immunological reaction between an antigen and an antibody (which selectively recognizes only this antigen).[32] ELISA satisfies the performance and validation criteria for screening methods and has been used for detection of many food and environmental toxic contaminants.[47] Application of ELISA in AA quantification requires synthesis of specific antibodies which was troublesome previously because of the low molecular mass of this compound. In general, substances with molecular mass below 1000 g mol^{-1} are not immunogenic and do not elicit the synthesis of antibodies.[32] The linkage between AA and the carrier protein as well as the choice of appropriate hapten are two key issues. According to Hu *et. al.*,[33] three approaches have been reported in principal. The first approach is using glutaraldehyde for conjugating while the second method is the most common method and uses active ester to combine AA and proteins, such as 1-ethyl-3-(3-dimethylaminopropyl)-carbodiimide (EDC) and *N*-hydroxysuccinimide (NHS). The third method uses *N*-acryloxysuccinimide (NAS) as the hapten. NAS contains the structures of AA and NHS, which makes it possible to react with the amino group and conjugate AA to the carrier protein directly.[33] Singh *et al.*,[48] used a method where polyclonal antibodies were raised against a hapten derived from AA and 3-mercaptobenzoic acid (3-MBA). According to their research, the assay was very specific to the AA-3-MBA derivative and showed no cross-reactivity to asparagine, AA, and 3-MBA. The assay was very sensitive with a limit of detection of 5.0 μg kg^{-1} in food models, and the results using this method were comparable with those obtained from MS methods including Food Analysis Performance Assessment Scheme control samples.[48] Compared with common analytical methods, ELISA is a method with similar sensitivity, lower cost, shorter detection time, simplicity, and does not requires expensive equipment or complex sample preparation. Therefore, ELISA kits available for commercial application are of high interest.[33] A novel bioanalytical technique that is used to quantify AA in foods involves electrochemical biosensors. In this method analytes are quantitatively determined by using highly selective bioreceptors whose function is based on the natural affinity of biologically active compounds for the substance of interest. Biosensors are a promising solution to AA quantification in foods because they offer many advantages over other methods, such as the relatively low price, high sensitivity and selectivity even in the case of samples characterized by a complex matrix composition, as well as the possibility of miniaturization which in turn enables on-line use.[32] Two methods are available for quantification of AA based on biosensors, which use either haemoglobin as a bio-component, or an oxygen electrode measuring the respiratory activity of microbial cells. A biosensor based on an amide-specific enzyme could be useful for quantification of AA in food, as well as in industrial effluents. Amidase from *Pseudomonas aeruginosa* exhibits transferase and hydrolase activities with aliphatic amide substrates, such as AA. This enzyme catalyses the transfer of acyl groups from amides to hydroxylamine forming acylhydroxamates and ammonia as well as the hydrolysis of aliphatic amides (*e.g.* AA) producing the corresponding acid and ammonia. Silva *et al.*,[49] used whole cells of *P. aeruginosa* containing amidase activity to develop a biosensor for toxic amides. It was successfully used for quantification of AA in real industrial effluents and recovery experiments were carried out which revealed an average substrate recovery of 93.3%. In most cases, AA is proposed to combine with a redox-active haemoglobin (Hb) [Hb-Fe(III) to Hb-Fe(II)] to form an adduct, which enlarges the distance between the redox centre of Hb and the electrode, suppressing the redox current. Efficient conjugation between AA, Hb, and the redox response of electrochemistry endow this method with high sensitivity, high selectivity, rapid response, and a wide linear range.[33] The biosensor is constructed using a carbon-paste electrode modified with Hb, which contains four prosthetic groups of heme-$Fe_{(III)}$. Such an electrode displays a reversible reduction/oxidation process of Hb-Fe(III)/Hb-Fe(II). The electrodes presented a very low detection limit equal to 1.2×10^{-10} M. The validation obtained in the matrix by water extraction of potato chips showed that the electrodes present are suitable for the direct determination of AA in food samples.[50] To harmonise the quality of analytical results, the method of analysis used for the analysis of AA content must comply with the performance criteria[27] shown in Table 2.3.

2.6 Occurrence of AA in Foods

Since 2002, the occurrence of AA in foods has been studied intensively as it has been found in starch and amino acid based, thermally treated products. While AA is not found in foods that are not fried or baked, it is found at very low levels in animal-based food products such as meat and fish that are boiled or microwaved.[51] Food groups containing higher levels of AA and that are important due to their relatively high consumption rate can be listed as follows: potato products (French fries, oven-baked chips, potato crisps, and so on), cereal-based foods such as breakfast cereals, cookies, biscuits, bread (especially toasted bread), pies, cakes, coffee, chicory, and

Table 2.3 Performance criteria of analytical methods used for determination of AA content, set by EC. Reproduced from ref. 27 with permission, Copyright © European Union, https://eur-lex.europa.eu, 1998–2019.

Parameter	Criterion
Applicability	Foods specified in EC Regulation[27]
Specificity	Free from matrix or spectral interferences
Field blanks	Less than limit of detection (LOD)
Repeatability (RSDr)	0.66 times RSDR as derived from (modified) Horwitz equation
Reproducibility (RSDR)	As derived from (modified) Horwitz equation
Recovery	75–110%
Limit of Detection (LOD)	Three tenths of LOQ
Limit of Quantification (LOQ)	For benchmark level <125 μg kg^{-1}: ≤two fifths of the benchmark level (however not required to be <20 μg kg^{-1}). For benchmark level ≥125 μg kg^{-1}: ≤50 μg kg^{-1}

other coffee substitutes, chocolate, teething biscuits, baby rusks, and various other baby foods.[44] The EFSA has proclaimed that roasted coffee and other cooked foods, for instance potatoes, crackers, cookies, toasted bread and breakfast cereals, are the major sources of AA, contributing more than 90% of the AA intake through our daily diets. The concentrations of AA in some food products, for instance potato chips and coffee, reach up to 3500 μg kg^{-1} or even higher than 4500 μg kg^{-1}.[53] Fried potato products are known to contain high levels of AA, which sometimes exceed 1000 μg kg^{-1}. This may be attributed to the high temperatures during processing and the high content of free asparagine in raw potatoes. However, Amrein *et al.* found that the content of free asparagine did not correlate with the potential for AA formation in potatoes, whereas a highly significant correlation between the content of reducing saccharides and AA was observed.[54] This is because the level of asparagine in raw potatoes varies only within a narrow range, whereas reducing saccharides content can vary widely depending on variety, storage condition, *etc.*[55] Esposito *et al.*[56] determined AA levels in potato crisps and other snacks. Based on their research, the concentration of AA detected ranged from 21 to 3444 μg kg^{-1} and the highest level occurred in potato crisp samples that showed a median value of 968 μg kg^{-1}. According to Capei *et al.*,[57] who conducted research on the determination AA levels in cereal-based foods, in particular biscuits and breakfast cereals, that the percentages of contaminated samples tested were very high (95.5% of the biscuits and 75% of the breakfast cereals) with a wide range of contamination: from 30 to 940 μg kg^{-1}. The maximum level of 940 μg kg^{-1} that was found in a sample was probably due to the industrial process conditions and to the presence of many different ingredients, such as wheat, oat, barley, rice, maize, rye. In the category of breakfast cereals tested, the occurrence of AA, ranging from 60 to 810 μg kg^{-1}, is less (75%) than that found in biscuits. Faraji *et al.*[58] collected data on the AA content in leavened wheat-rye breads and wheat rolls in order to demonstrate the influence of various technological factors on their contents. The levels of AA in a common type of bread baked in various bakeries were relatively low, ranging from 7 to 22 μg kg^{-1}. Varying the content of rye flour did not seem to have any impact on AA levels. Lower AA contents were found in breads leavened by natural rye sourdough, probably due to the lower pH values of dough achieved during fermentation. It can be concluded that the content of AA in breads is relatively low, however, these findings were in accordance with the EFSA report,[6] which identified soft breads, unlike crisp breads, as a food category with a low AA content. AA levels in roasted coffee are affected by concomitant formation and reduction reactions during the roasting process. AA formation reactions are dominant at the beginning of the roasting cycle, leading to increased levels at this stage, and then decline steeply toward the end of the roasting cycle due to higher rates of elimination (physical and chemical loss). Therefore, light-roasted coffees may contain relatively higher amounts of AA than very dark-roasted beans.[59] Khan *et al.*[53] investigated the contents of AA in 56 food samples including Arabic coffee '*Qahwa*', and coffee and tea of different brands and origin. AA was detected in all of the analysed coffee and the concentrations ranged from 10 to 682 μg kg^{-1}. In comparison to other coffees, the Arabic coffee '*Qahwa*' and tea contained lower amounts of AA. Samples obtained from four classes of tea products including black tea, oolong tea, white tea and yellow tea had quite a low AA concentration (<20 μg kg^{-1}) probably because the fresh leaves were not dried at high temperature. High AA levels occurred in baked, roasted, and one sun-dried green tea sample (46–94 μg kg^{-1}), probably because 100–150 °C are the usual temperatures used during processing.[60] Indicative findings of AA content in foods are shown in Table 2.4.

2.7 Mitigation of AA Content in Foods

A major challenge lies in developing an effective strategy for AA mitigation in foods without altering the nutritional and organoleptic properties, while considering safety issues. Food scientists around the world have developed various methods to mitigate the AA content in thermally treated food products. Mitigation techniques using additives such as salts, amino acids, cations and organic acids along with blanching of foods have considerably reduced the AA content. The use of secondary metabolites such as polyphenols, antioxidants and spices (*e.g.* cardamom, white pepper) also affects AA content in fried food products.[80] Other mitigation techniques such as asparaginase pre-treatment and low-temperature air frying with chitosan have been effective in mitigating the concentration of AA. A combined pre-treatment process along with the use of additives is the latest trend in AA mitigation.[67] Mitigation strategies should be applied permanently; any levels that exceed the BLs of AA in foods should initiate a systematic approach by the food producer in the strategy of mitigation of AA content in its products. BLs for the presence of AA in foods are shown in Table 2.5.

Table 2.4 Benchmark levels for the presence of AA in foods set by European Commission. Reproduced from ref. 27 with permission, Copyright © European Union, https://eur-lex.europa.eu, 1998–2019.

Food	Benchmark level (μg kg^{-1})
French fries (ready-to-eat)	500
Potato crisps from fresh potatoes and from potato dough	750
Potato-based crackers	
Other potato products from potato dough	
Soft bread	
a) Wheatbased bread	50
b) Soft bread other than wheatbased bread	100
Breakfast cereals (excluding porridge)	
- bran products and whole grain cereals, gun puffed grain	300
- wheat and rye based products[a]	300
- maize, oat, spelt, barley and rice-based products[b]	150
Biscuits and wafers	350
Crackers with the exception of potato-based crackers	400
Crispbread	350
Ginger bread	800
Products similar to the other products in this category	300
Roast coffee	400
Instant (soluble) coffee	850
Coffee substitutes	
a) coffee substitutes exclusively from cereals	500
b) coffee substitutes from a mixture of cereals and chicory	[b]
c) coffee substitutes exclusively from chicory	4000
Baby foods, processed cereal-based foods for infants and young Children excluding biscuits and rusks[c]	50
Biscuits and rusks for infants and young children[c]	150

[a]Non-whole grain and/or non-bran based cereals. The cereal present in the largest quantity determines the category.
[b]The benchmark level to be applied to coffee substitutes from a mixture of cereals and chicory takes into account the relative proportion of these ingredients in the final product.
[c]As defined in Regulation (EU) No 609/2013.

With regard to raw materials/foods production, mitigation processes can be categorised as:

- production of raw materials with decreased content of precursors;
- elimination of precursors during technological treatments;
- elimination of AA by stimulation of its polymerisation reactions.

2.7.1 Production of Raw Material with Decreased Content of Precursors

Reducing saccharides are, in general, considered as the limiting factor in the formation of AA, because asparagine is often far in excess of reducing saccharides in potato tubers, and a significant correlation exists between

Table 2.5 AA content found in a range of foods.

Food product	n (Number of samples)	Mean (μg kg^{-1})	Acrylamide range (μg kg^{-1})	Reference
Potatoes (raw)	—	—	<10–<50	41 and 51
Potato chips/crisps	—	—	117–4215	41 and 51
	30	1162	173–3444	56
	17	664.18	186.42–1828.75	39
Corn-based extruded snacks	20	257	123–447	56
Savory snacks	20	258	44–813	56
French fries/chips	97	—	20–2762	30
	—	—	59–5200	41 and 51
Bakery products and biscuits	—	—	18–3324	41 and 51
	17	297.18	30–940	57
Breads	—	—	<10–3200	41 and 51
	12	16	7–22	58
	49	—	0–432	30
Breakfast cereals	5	263.75	60–810	57
	—	—	<10–1649	41 and 51
	23	—	11–1057	30
Chocolate	—	—	<2–826	41 and 51
	14	—	0–909	30
Roasted coffee	—	—	45–935	41 and 51
	59	—	37–539	30
	25	371.60	152–682	53
Coffee substitute	—	—	80–5399	41 and 51
Green tea	4	15.50	10–23	53
	10	48.43	0–94	60
Black tea	17	77.53	62–97	53
Dried prunes	13	60.97	14.74–124.26	61
Roasted peanuts	5	21	11–37	56
	26	15.74	6.16–42.86	61
Baby food and infant formula	—	—	<10–130	41 and 51
	36	—	0–130	30
	10	0.60	0.0–2.0	52
Grilled beef and chicken	16	70.69	23–250	64
Seafood	14	20.11	5.58–50.35	66

reducing saccharides content and AA content.[68–70] Cultivars with low reducing saccharide content usually generate less AA in fried potato products, suggesting that cultivar selection would be an effective approach to decrease AA content in thermally treated products.[15] Nitrogen (N) management is another practice that can influence saccharide content and AA-forming potential in fried potato products. Plants with adequate N supply produce tubers with low reducing saccharides content at harvest and through storage.[71,72]

Storage conditions strongly affect reducing sugar concentrations. Temperatures greater than 8 °C will reduce cold-induced sweetening during storage of potato tubers. However, higher temperatures promote aging and may induce senescent sweetening.[27,73] According to EC regulation,[27] food producers should identify and use the potato varieties that are not only suitable for the product type, but also where the content of AA precursors, such as reducing saccharides (fructose and glucose) and asparagine content, is the lowest as much as possible. The stored potatoes should be used within their optimal storage window and potato varieties with lower AA-forming potential in cultivation, storage and during food processing should be identified. In defined terms it means:

- the temperature should be appropriate to the potato variety stored and it should be above 6 °C;
- the level of humidity should be such as to minimise senescent sweetening;
- sprouting should be suppressed in long-term stored potatoes where permitted, using appropriate agents;
- during storage the level of reducing saccharides in potatoes should be tested.

2.7.2 Elimination of Precursors During Technological Treatments

In principal, there are two possible ways to eliminate precursors, either application of L-asparaginase able to hydrolyse asparagine to aspartic acid, which is not able to provide AA, or use of glucose-fructose oxidoreductase to form D-glucono-delta-lactone and D-glucitol, respectively. Considering the real contents of precursors in a food matrix it is clear that elimination of L-asparagine is more reasonable in comparison to glucose–fructose elimination. Therefore, L-asparaginase has become a powerful tool for AA mitigation in the food industry. With the success of commercial products, it is likely that asparaginase will be used more and more. The first commercially available "acrylamide-free" product, biscuits treated with Preventase™ (commercial form of L-asparaginase), was announced for launch in Germany 2008. However, potential adverse effects of L-asparaginase treatment on the sensory properties of cooked foods and the need to achieve sufficient enzyme–substrate contact are areas for future research. So, if the application of asparaginase becomes commercially attractive, its use alongside raw materials low in asparagine may provide the solution to the AA problem.[74]

With regard to application in a potato matrix, different amounts of L-asparaginase at two temperatures (20 °C and 37 °C) were compared in a system simulating the composition of raw potatoes and subsequently in fresh and stored potatoes, respectively, and also in dried potato products. It was found that the addition of as little as 0.2 U g^{-1} of lyophilized enzyme per 1 g of fresh potato mash resulted in an approximately 50% reduction in AA content. A higher amount of enzyme, 1 U g^{-1}, and higher incubation temperature (37 °C) of fresh potato sample with the enzyme produced a greater level of suppression of AA production (up to 97%) when exposed to the same heat regime (180 °C, 20 min). In the case of potatoes stored at 4 °C, AA content four times higher than in fresh potatoes was observed. The addition of enzyme equal to 2 U g^{-1} and subsequent incubation of the mixture at 37 °C for 30 min led to a 70% reduction in AA content. Almost total mitigation of AA content (90–97%) was achieved in products prepared from dried potato powder treated with L-asparaginase.[75,76] On the other hand, application of L-asparaginase may have limited or no effect on the levels of AA in products with high fat content, low moisture or high pH value.[27]

2.7.3 Elimination of AA by Stimulation of Its Polymerisation Reactions

As follows from the physico-chemical characteristics and molecular structure shown in Scheme 2.1, AA is potentially a reactive compound that provides an opportunity for its mitigation on the basis of its interactions with other molecules, or even with its molecules themselves. When a mixture of L-asparagine and D-glucose was heated in the presence of NaCl at 171.17 °C for 10 min, the AA content was lowered by approximately 40% in comparison to a mixture without NaCl addition. It should be noted that effective lowering of AA content was observed even at 1% NaCl addition, which could make possible the application of this knowledge in real technological procedures of food processing where the current level of NaCl addition lies at around 1.5%.[77] Differential scanning calorimetry (DSC) can provide a very illustrative view of AA behaviour during heating, especially in the presence of other compounds such as NaCl.[78] DSC records of AA, NaCl and a mixture of AA : NaCl 1 : 1 (w/w) are shown in Figure 2.5, where NaCl is thermally inert at the chosen temperature interval.

However, the DSC record of pure AA exhibits two different peaks. The first endothermic peak with an onset temperature at 83 °C corresponds to the melting point of AA. The second exothermic peak, with an onset at 156 °C, obviously corresponds to the polymerization reaction. The DSC record of the AA/NaCl mixture shows the melting peak of AA. However, the start of the second exothermic peak is noticeably shifted down to 144 °C. As follows from Figure 2.5, NaCl decreases the starting temperature of AA polymerization by 12 °C in comparison with pure AA, which indicates a strong acceleration of the polymerization process. However, when AA was applied onto NaCl and two additives of table salt – potassium ferrocyanide and potassium iodate, the amount of AA applied onto NaCl decreased by only 13%, whereas the amounts of AA applied onto potassium ferrocyanide and potassium iodate decreased much more considerably – by 61% and 88%, respectively.[79] The decrease in the amounts of AA is shown in Figure 2.6.

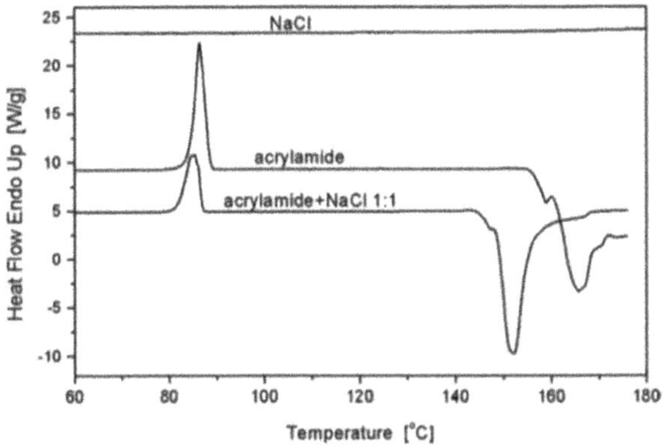

Figure 2.5 DSC records of AA, NaCl, and a mixture of AA and NaCl. Reproduced from ref. 78 with permission from John Wiley and Sons, Copyright © 2007 Institute of Food Technologists.

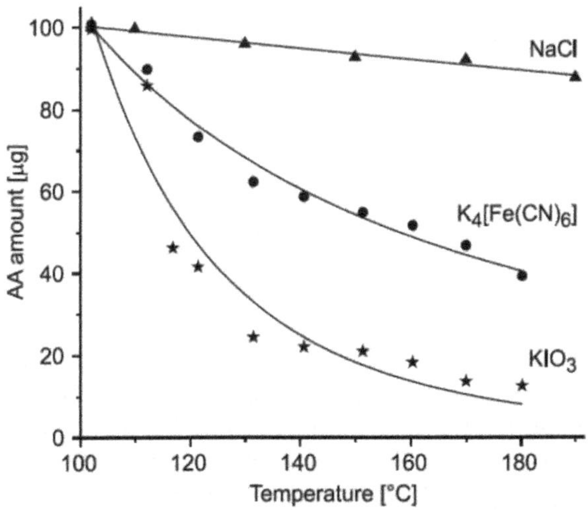

Figure 2.6 Decrease in the amounts of AA when heated in the presence of NaCl, $K_4[Fe(CN)_6]$, and KIO_3. Reproduced from ref. 79 [available at http://www.vup.sk/download.php?bulID=35] with permission, Copyright © 2007 VUP Food Research Institute, Bratislava.

From a practical point of view, it is very important that both compounds are used frequently in table salt production as additives. $K_4[Fe(CN)_6]$ is the compound permitted for usage in the food industry and its number within the international numbering system is INS 535. The compound has the same number in European food legislation, and it is being added to table salt at levels up to 20 mg kg^{-1} as an anticaking agent since it adsorbs water and so prevents the table salt from forming clumps and being difficult to pour. In some countries, KIO_3 is used for iodination of table salt at levels up to

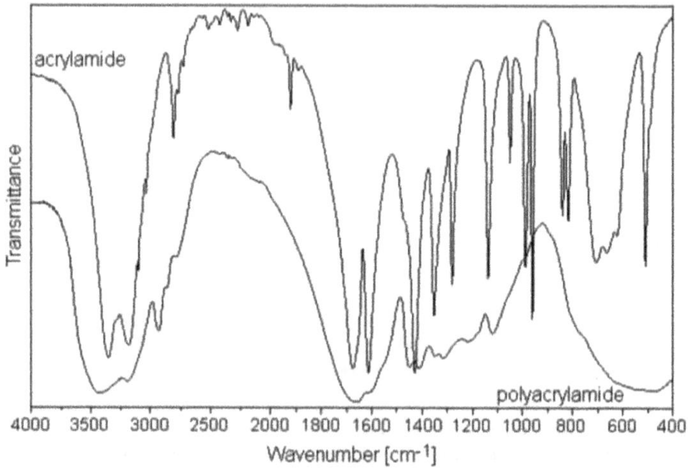

Figure 2.7 Infra-red spectrum of AA (top) and its polymer (bottom) measured in a KBr pellet at room temperature. Reproduced from ref. 79 [available at http://www.vup.sk/download.php?bulID=35] with permission, Copyright © 2007 VUP Food Research Institute, Bratislava.

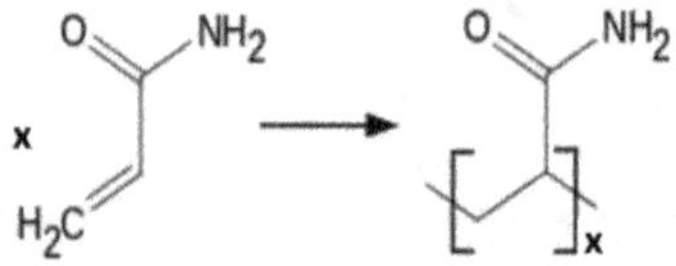

Scheme 2.3

35 mg kg^{-1} instead of KI, because iodide can be oxidized to iodine under wet conditions by oxygen with subsequent sublimation out of the salt. Moreover, KIO_3 is occasionally used as a maturing agent in baking technology. As these results show, both compounds could strengthen considerably the effect of NaCl on the elimination of AA in a real food matrix. Comparing infrared spectra of pure AA and the product formed during the experiments (shown in Figure 2.7).

It was found that all the salts under study brought about the polymerization of AA through the formation of a polymer with a C–C backbone, whereas the carbonyl and amino groups remained unchanged. The polymerisation reaction is shown in Scheme 2.3.

So, potassium ferrocyanide and potassium iodate exhibit a much stronger effect on AA polymerization than pure sodium chloride itself. For this reason, they could strengthen considerably the efficiency of AA elimination in a real food matrix with regard to their presence in table salt or being added directly to the thermally treated food matrix, respectively. Overall, the polymerization of AA may represent one conceivable pathway of AA elimination in a real food matrix.

2.8 Conclusion

AA is a typical food processing contaminant formed during the early stage of the Maillard reaction from its precursors such as reducing saccharides and the amino acid asparagine during heating of raw materials at over 120 °C. Due to its ability to form DNA adducts *via* GA, AA is therefore listed by the IARC as a Group 2A toxicant and probably carcinogenic in humans. AA content has frequently been studied from the points of its formation and occurrence in heat-treated foods to find effective solutions for the current situation. On the basis of EFSA data, the EC set out indicative contents, so called BLs, serving as indicators of the level of food contamination by AA. An important part of mitigation strategies lies in preventative steps to decrease precursors in raw materials, degrade precursors during technological treatments, or decrease the level of AA content by stimulation of its polymerisation reactions leading to formation of the harmless compound polyacrylamide.

Acknowledgement

This contribution is the result of the project implementation "Building Infrastructure for Modern Research of Civilization's Diseases" (ITMS 26230120009) financially supported by the Research & Development Operational Programme funded by the ERDF, grant VEGA 1/0487/16 and APVV- 061-2018.

References

1. P. Šimko, in *Advances in Food and Nutrition Research*, ed. F. Toldrá, Elsevier, London, 2018, vol. 84, p. 241.
2. E. Tareke, P. Rydberg, P. Karlsson, S. Eriksson and M. Törnqvist, *J. Agric. Food Chem.*, 2002, **50**, 4998.
3. R. H. Stadler, I. Blank, N. Varga, F. Robert, J. Hau, P. A. Guy, M. C. Robert and S. Riediker, *Nature*, 2002, **419**, 449.
4. D. S. Mottram, B. L. Wedzicha and A. T. Dodson, *Nature*, 2002, **419**, 448.
5. J. E. Hodge, *J. Agric. Food Chem.*, 1953, **1**, 928.
6. EFSA Panel on Contaminants in the Food Chain, *EFSA J.*, 2015, **13**, 24.
7. M. Biedermann, S. Biedermann-Brem, A. Noti and K. Grob, *Mitt. Lebensmittelunters. Hyg.*, 2002, **93**, 653.
8. A. Becalski, B. Brady, S. Feng, B. R. Gauthier and T. Zhao, *Food Addit. Contam.*, 2011, **28**, 726.
9. D. V. Zyzak, R. A. Sanders, M. Stojanovic, D. H. Tallmadge, B. L. Eberhart, D. K. Ewald, D. C. Gruber, T. R. Morsch, M. A. Strothers, G. P. Rizzi and M. D. Villagran, *J. Agric. Food Chem.*, 2003, **51**, 4782.
10. M. Granvogl, M. Jezussek, P. Koehler and P. Schieberle, *J. Agric. Food Chem.*, 2004, **52**, 4751.

11. R. Stadler and G. Scholz, Acrylamide, *Nutr. Rev.*, 2004, **62**, 449.
12. A. Claus, G. M. Weisz, A. Schieber and R. Carle, *Mol. Nutr. Food Res.*, 2006, **50**, 87.
13. F. J. Casado, A. Montaño, D. Spitzner and R. Carle, Investigations into acrylamide precursors in sterilized table olives: Evidence of a peptic fraction being responsible for acrylamide formation, *Food Chem.*, 2013, **141**, 1158.
14. M. Granvogl, P. Koehler, L. Latzer and P. Schieberle, *J. Agric. Food Chem.*, 2008, **56**, 6087.
15. N. G. Halford, T. Y. Curtis, N. Muttucumaru, J. Postles, J. S. Elmore and D. S. Mottram, *J. Exp. Bot.*, 2012, **63**, 2841.
16. A. Adams, S. Hamdani, F. Van Lancker, S. Mejri and N. De Kimpe, *Food Res. Int.*, 2010, **43**, 1517.
17. V. A. Yaylayan, A. Wnorowski and C. Perez-Locas, *J. Agric. Food Chem.*, 2003, **51**, 1753.
18. V. A. Yaylayan, *J. Food Nutr. Res.*, 2009, **48**(1), 1.
19. F. Robert, G. Vuataz, P. Pollien, F. Saucy, M. I. Alonso, I. Bauwens and I. Blank, *J. Agric. Food Chem.*, 2004, **52**, 6837.
20. I. Blank, *Ann. N. Y. Acad. Sci.*, 2005, **1043**, 30.
21. International Agency for Research on Cancer, *Monographs on the Evaluations of Carcinogenic Risks to Humans*, 1994, vol. 60, p. 389.
22. EFSA Panel on Contaminants in the Food Chain, *EFSA J.*, 2015, **13**, 14.
23. Council Regulation (EEC) No 315/93, *Off. J. Eur. Communities: Legis.*, 1993, **37**, 1.
24. Commission Regulation (EC) 1881/2006/EC, *Off. J. Eur. Communities: Legis.*, 2006, **364**, 5.
25. Commission Recommendation 2010/307/EU, *Off. J. Eur. Communities: Legis.*, 2010, **137**, 4.
26. Commission Recommendation 2007/331/EC, *Off. J. Eur. Communities: Legis.*, 2007, **123**, 33.
27. Commission Regulation (EC) 2017/2158, *Off. J. Eur. Communities: Legis.*, 2017, **304**, 24.
28. E. Tareke, P. Rydberg, P. Karlsson, S. Eriksson and M. Törnqvist, *Chem. Res. Toxicol.*, 2000, **13**(6), 517.
29. E. V. Petersson, J. Rosén, C. Turner, R. Danielsson and K.-E. Hellenäs, *Anal. Chim. Acta*, 2006, **557**(1–2), 287.
30. E. Dybing, P. B. Farmer, M. Andersen, T. R. Fennell, S. P. D. Lalljie, D. J. G. Müller, S. Olin, B. J. Petersen, J. Schlatter, G. Scholz, J. A. Scimeca, N. Slimani, M. Tornqvist, S. Tuijtelaars and P. Verger, *Food Chem. Toxicol.*, 2005, **43**(3), 365.
31. S. H. Kim, J. H. Hwang and K. G. Lee, *Food Sci. Biotechnol.*, 2011, **20**(3), 835.
32. J. Oracz, E. Nebesny and D. Żyżelewicz, *Talanta*, 2011, **86**, 23.
33. Q. Hu, X. Xu, Y. Fu and Y. Li, *Food Control*, 2015, **56**, 135.
34. Y. Zhu, G. Li, Y. Duan, S. Chen, C. Zhang and Y. Li, *Food Chem.*, 2008, **109**(4), 899.

35. A. A. Elbashir, M. M. A. Omar, W. A. W. Ibrahim, O. J. Schmitz and H. Y. Aboul-Enein, *Crit. Rev. Anal. Chem.*, 2014, **44**(2), 107.
36. Y. Zhang, G. Zhang and Y. Zhang, *J. Chromatogr. A*, 2005, **1075**(1–2), 1.
37. F. Khoshnam, B. Zargar, N. Pourreza and H. Parham, *J. Iran. Chem. Soc.*, 2010, **7**(4), 853.
38. J. Keramat, A. LeBail, C. Prost and N. Soltanizadeh, *Food Bioprocess Technol.*, 2011, **4**(3), 340.
39. J. O. Fernandes and C. Soares, *J. Chromatogr. A*, 2007, **1175**(1), 1.
40. M. M. A. Omar, A. A. Elbashir and O. J. Schmitz, *Food Chem.*, 2015, **176**, 342.
41. D. R. Lineback, J. R. Coughlin and R. H. Stadler, *Annu. Rev. Food Sci. Technol.*, 2012, **3**, 15.
42. H. Wang, A. W. M. Lee, S. Shuang and M. M. F. Choi, *Microchem. J.*, 2008, **89**(2), 90.
43. H. Wang, F. Feng, Y. Guo, S. Shuang and M. M. F. Choi, *J. Food Compos. Anal.*, 2013, **31**(1), 7.
44. S. E. K. Tekkeli, C. Önal and A. Önal, *Food Anal. Methods*, 2012, **5**(1), 29.
45. M. A. Majcher and H. H. Jeleń, *Food Addit. Contam.*, 2007, **24**(4), 337.
46. F. Pedreschi, P. Moyano, K. Kaack and K. Granby, *Food Res. Int.*, 2005, **38**(1), 1.
47. M. Franek, D. Rubio, I. Diblikova and F. Rubio, *Talanta*, 2014, **123**, 146.
48. G. Singh, B. Brady, T. Koerner, A. Becalski, T. Zhao, S. Feng, S. B. Godefroy, A. C. Huet and P. Delahaut, *Food Anal. Methods*, 2014, **7**(6), 1298.
49. N. Silva, D. Gil, A. Karmali and M. Matos, *Biocatal. Biotransform.*, 2009, **27**(2), 143.
50. A. Stobiecka, H. Radecka and J. Radecki, *Biosens. Bioelectron.*, 2007, **22**(9–10), 2165.
51. T. Krishnakumar and R. Visvanathan, *J. Food Process. Technol.*, 2014, **5**(7), 1.
52. M. Lambert, C. Inthavong, F. Hommet, J.-C. Leblanc, M. Hulin and T. Guérin, *Food Chem.*, 2018, **240**, 997.
53. M. R. Khan, Z. A. Alothman, M. Naushad, A. K. Alomary, S. M. Alfadul, I. H. Alsohaimi and M. S. Algamdi, *Sci. Rep.*, 2017, **7**, 1.
54. T. M. Amrein, L. Andres, F. Escher and R. Omadò, *Food Addit. Contam.*, 2007, **24**(1), 13.
55. S. Romani, M. Bacchiocca, P. Rocculi and M. Dalla Rosa, *J. Food Compos. Anal.*, 2009, **22**(6), 582.
56. F. Esposito, A. Nardone, E. Fasano, M. Triassi and T. Cirillo, *Food Chem. Toxicol.*, 2017, **108**, 249.
57. R. Capei, L. Pettini, A. Lo Nostro and G. Pesavento, *J. Prev. Med. Hyg.*, 2015, **56**(4), 190.
58. M. Faraji, M. Hamdamali, F. Aryanasab and M. Shabanian, *J. Chromatogr. A*, 2018, **1558**, 14.
59. H. Guenther, E. Anklam, T. Wenzl and R. H. Stadler, *Food Addit. Contam.*, 2007, **24**(1), 60.

60. J. Liu, G. Zhao, Y. Yuan, F. Chen and X. Hu, *Food Chem.*, 2008, **108**(2), 760.
61. E. L. De Paola, G. Montevecchi, F. Masino, D. Garbini, M. Barbarena and A. Antonelli, *Food Chem.*, 2017, **217**, 191.
62. J. S. Elmore, A. Briddon, A. T. Dodson, N. Muttucumaru, N. G. Halford and D. S. Mottram, *Food Chem.*, 2015, **182**, 1.
63. J. Michalak, E. Gujska and A. Kuncewicz, *J. Food Compos. Anal.*, 2013, **32**(1), 68.
64. O. Kaplan, G. Kaya, C. Ozcan, M. Ince and M. Yaman, *Microchem. J.*, 2009, **93**(2), 173.
65. M. Surma, A. Sadowska-Rociek, E. Cieślik and K. Sznajder-Katarzyńska, *Microchem. J.*, 2017, **131**, 98.
66. L. Qin, Y. Y. Zhang, X.-B. Xu, X. S. Wang, H. W. Liu, D. Y. Zhou, B. W. Zhu and M. Thornton, *Food Chem.*, 2017, **232**, 633.
67. G. Baskar and R. Aiswarya, *J. Sci. Food Agric.*, 2018, 4385–4394.
68. T. M. Amrein, B. Schönbächler, F. Rohner, H. Lukac, H. Schneider, A. Keiser, F. Escher and R. Amadò, *Eur. Food Res. Technol.*, 2004, **219**, 572.
69. A. Ohara-Takada, C. Matsuura-Endo, Y. Chuda, H. Ono, H. Yada, M. Yoshida, A. Kobayashi, S. Tsuda, S. Takigawa, T. Noda and H. Yamauchi, *Biosci., Biotechnol., Biochem.*, 2005, **69**, 1232.
70. J. S. Elmore, A. Briddon, A. T. Dodson, N. Muttucumaru, N. G. Halford and D. S. Mottram, *Food Chem.*, 2015, **182**, 1.
71. D. Kumar, B. P. Singh and P. Kumar, *Ann. Appl. Biol.*, 2004, **145**, 247.
72. N. Sun, C. J. Rosen and A. L. Thompson, *Am. J. Potato Res.*, 2018, **95**(5), 473.
73. C. Rosen, N. Sun, N. Olsen, M. Thornton, M. Pavek, L. Knowles and R. N. Knowles, *Am. J. Potato Res.*, 2018, **95**(4), 319.
74. F. Xu, M. J. Oruna-Concha and J. S. Elmore, *Food Chem.*, 2016, **210**, 163.
75. Z. Ciesarová, E. Kiss and P. Boegl, *J. Food Nutr. Res.*, 2006, **45**(4), 141.
76. Z. Ciesarová, K. Kukurová and C. Benešová, *Nutr. Food Sci.*, 2010, **40**(1), 55.
77. E. Kolek, P. Šimko and P. Šimon, *Eur. Food Res. Technol.*, 2006, **224**, 283.
78. E. Kolek, P. Šimon and P. Šimko, *J. Food Sci.*, 2007, **73**(1), 341.
79. E. Kolek, P. Šimko, P. Šimon and A. Gatial, *J. Food Nutr. Res.*, 2007, **46**(1), 39.
80. L. Marková, Z. Ciesarová, K. Kukurová, H. Zieliňski, M. Przygodzka, A. Bednáriková and P. Šimko, *Chem. Pap.*, 2012, **66**, 949.

CHAPTER 3

Risk/Benefits Evaluation of Acrylamide Mitigation Initiatives in Cereal Products

MARTA MESÍAS, CRISTINA DELGADO-ANDRADE AND FRANCISCO J. MORALES*

Institute of Food Science, Technology and Nutrition (ICTAN-CSIC), Jose Antonio Novais 10, 28040 Madrid, Spain
*E-mail: fjmorales@ictan.csic.es

3.1 Food Contaminants: Acrylamide

Food contaminants comprise a wide group of compounds that are undesirable because they can have an adverse effect on human health. Food contaminants may be present as a result of production (including management of crops, pre-/post-harvest practices, animals, *etc.*), preparation, food formulation, processing, packaging, transport and storage, or as a result of environmental contamination. Processing contaminants are defined as compounds produced in a food when it is cooked or processed, but they are not present or are present at much lower concentrations in the raw or the unprocessed food recipe.[1] Unlike with the classic concept of a contaminant, process contaminants are not introduced accidentally from an external source or as result of microbial, fungal, animal activity, or from packaging or the

Food Chemistry, Function and Analysis No.19
Mitigating Contamination from Food Processing
Edited by Catherine S. Birch and Graham A. Bonwick
© The Royal Society of Chemistry 2020
Published by the Royal Society of Chemistry, www.rsc.org

environment. Particularly, heat-generated process contaminants, also called thermal-generated process contaminants, form after thermal/heat treatment of foods. Chemical contaminants generated that form in this group are acrylamide, furan, heterocyclic aromatic amines, acrolein, 3-monochloropropanodiol esters and glycidol esters, among others.

Acrylamide in cereal-based foods is formed from the free amino acid asparagine in the presence of reducing sugars (glucose, fructose, and maltose) or carbonyl compounds through the Maillard reaction. These acrylamide precursors occur naturally in plant materials including, cereal grains, potatoes, cocoa, and coffee. During the initial stage of the Maillard reaction, the α-amine group of free asparagine reacts with a carbonyl source, forming a Schiff base. Indeed, the side chain amide group of asparagine is incorporated into the amide bond of acrylamide. The decarboxylated Amadori product of asparagine and its subsequent β-elimination of an imine will efficiently release acrylamide.[2] In this reaction pathway, 3-aminopropionamide has been suggested as a transient intermediate. 3-Aminopropionamide is also formed in foods by an enzymatic decarboxylation of asparagine.[3] As a minor pathway, in the case of cereals and particularly for wheat flour, a peptide sequence containing a β-proton in the amino acid adjacent to alanine of the gluten structure may fragment in a pyrolytic pathway under common baking conditions and release acrylamide.[4] There is also evidence that asparagine itself may be converted thermally into acrylamide through decarboxylation and deamination reactions,[5] but the yield of acrylamide is much higher when a compound with a carbonyl moiety is present.[5] The length of the carbonyl compound is relevant since a shorter carbonyl molecule is strained to form a cyclic hemiacetal structure and subsequently the carbonyl becomes more readily available for nucleophilic attack from the alpha-amine of asparagine.[6] The reactivity towards acrylamide formation increases as D-glucose < 2-deoxyglucose < ribose < glyceraldehyde < glyoxal. However, dicarbonyl compounds are not essential for the formation of acrylamide from asparagine. The reactivity of fructose is nearly 10 times that of glucose and 50 times that of sucrose. Sucrose can also participate by previous enzymatic hydrolysis or by an acid-catalyzed reaction. Maltose reactivity to form acrylamide is similar to that of glucose.[7]

The Maillard reaction is a rather complex cascade of reactions, generating a plethora of desirable flavour, aroma and colour compounds when cereal-based foods are processed. Hence, the Maillard reaction improves food palatability and the sensorial properties of cereal-based foods that impact positively on consumer acceptance. Since acrylamide formation in cereal-based foods is concomitant with the extent of the Maillard reaction, any strategy to reduce or prevent the formation of acrylamide should aim to ensure that beneficial attributes of the cooked foods are not adversely affected. The beneficial and potential risks of acrylamide mitigation strategies in bakery products are summarized in Figure 3.1.

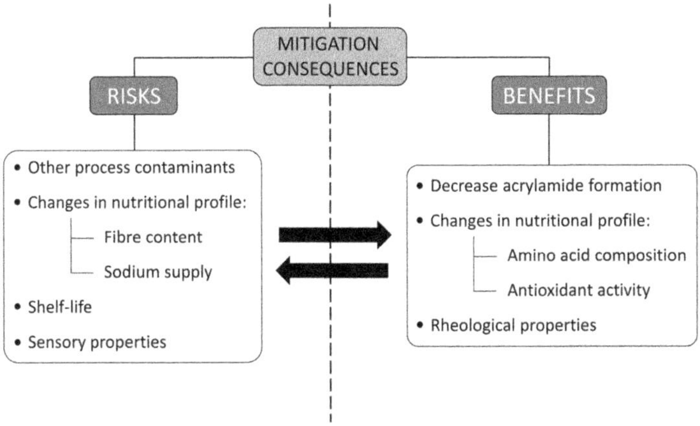

Figure 3.1 Consequences of acrylamide mitigation strategies in bakery products.

3.2 Precursors of Acrylamide in Cereals

Asparagine has been identified as the key factor in the formation of acrylamide in cereal foods, since the sugar content is in excess, between 10 and 50 times higher than that of asparagine.[8] Sugar concentrations in cereal grain, as a percentage of total dry matter, are approximately 1% glucose, 0.2% fructose, 0.5% maltose, and 0.5% sucrose.[9] Consequently, reduction of the asparagine content in the flour is the primary strategy to prevent the formation of acrylamide during processing of cereals, and particularly during baking. However, the free asparagine content differs greatly among cereals aimed at human consumption. The main cereals for human consumption are wheat (*Triticum aestivum*), rice (*Oryza sativa*), maize (*Zea mays*), oat (*Avena sativa*), rye (*Secale cereale*), and spelt (*Triticum spelta*), in addition to barley (*Hordeum vulgare*), millet and sorghum (*Sorghum bicolor* L.). Wheat and rice are the major grains with the largest worldwide production, and thirdly maize. The minor grains are barley, sorghum, millet, oats and rye. In a fundamental report from Curtis *et al.*[10] they concluded that information on free asparagine concentration should be available to the secondary transformation cereal industry when a new variety of cereal is launched. These researchers identified eight varieties from the 73 investigated in their studies that have shown low free asparagine content in trials at different locations in the UK over several years. These varieties were Claire, Cocoon, Cordiale, Croft, Delphi, Horatio, Monterey and Myriad. Nevertheless, the solution to the problem is not simple since extensive monitoring of the free asparagine content in European wheat varieties reveals at least a two-fold range within the same crop year, due to changing environmental conditions, and up to five-fold variation between varieties.[11] Free asparagine concentrations ranged from 0.6 to 4.4 mmol kg^{-1} in wheat crops.[12] Despite similar agronomic practice being followed at each field location around the UK, significant variance arose from location and harvest year.[12] In addition, asparagine content in different cereal species and varieties within species vary widely. Asparagine, aspartic acid, and glutamic acid were the most abundant free amino acids, and asparagine comprised 30–50% of the total.[8] A rough classification of cereals according to their asparagine content is as follows: rice < maize < barley < spelt < wheat < oat < rye. The asparagine concentration in rye flour is about 634 mg kg^{-1} compared with wheat flour, 172 mg kg^{-1},[13] oat, 250–400 mg kg^{-1}, maize, 50–70 mg kg^{-1}, rice, 1–6 mg kg^{-1}, or spelt flour, 60–120 mg kg^{-1}.[7,14] Rye has the highest content of asparagine, about four-fold that of wheat. Additionally, asparagine represents 26% of the total free amino acid content in rye.[15] Bran derived from oat, and especially from wheat, has higher asparagine concentration than the corresponding flour from these other cereal species.[9]

Asparagine cannot be easily eliminated since it is an important amino acid in the nitrogen-storage cycle of plants, and of course of grains as well.[11] Asparagine accumulates in all plants during normal physiological processes such as seed germination and nitrogen transport in response to external stresses caused by drought, salt stress, mineral deficiencies, presence of toxic metals and pathogen attacks.[15] In this respect, the mitigation of acrylamide in cereal foods derived from the raw material is a challenge, and one strategy is the promotion of breeding programs to produce new crop varieties that harbor naturally lower amounts of free asparagine, that will reduce the risk of acrylamide formation.

Additionally, asparagine is not homogenously distributed in the grain.[15] The cereal grain is constituted by the endosperm (representing approximately 80% of the grain by weight), the bran and the germ. The asparagine is concentrated in the germ and the degree of milling will impact on the asparagine content of the cereal (Table 3.1). Flour extraction rates are usually used to express the portion of the endosperm that is separated into a particular grade of flour. The extraction rate will range from no reduction in wholemeal flours (100% extraction that contains all fractions of the grain) to the highest reduction in white flours (reaching approximately 70% extraction). Wholegrain products have higher asparagine concentrations than refined grains, and produce higher levels of acrylamide in finished products.

Nowadays consumers are aware of what they eat, selecting healthier foods, especially in terms of healthy diets and disease prevention, and wholegrain

Table 3.1 Proximate levels of acrylamide precursors (including sucrose) in wheat with a 14% of total protein content.

	Fraction (%)	Asparagine (mg kg^{-1})	Fructose (mg kg^{-1})	Glucose (mg kg^{-1})	Maltose (mg kg^{-1})	Sucrose (mg kg^{-1})
Whole grain	100	374	107	163	5940	5160
Flour	82	141	400	600	8900	410
Bran	15	911	8060	7760	48	9450
Germ	3	1506	4810	4110	5050	85 272

foods are a preferential choice. According to Englyst *et al.*,[16] dietary fiber can be categorized into soluble dietary fiber (*i.e.* pectin) which is digested into carbohydrate and absorbed in the small intestine, and insoluble dietary fiber (*i.e.* cellulose, hemicellulose, lignin *etc.*), which is unable to be digested and is poorly metabolized in the small intestine. Insoluble dietary fiber exerts physiological health benefits in humans, including a reduction of postprandial blood glucose, pre-prandial cholesterol levels, coronary heart disease, and others.[17] Dietary fiber content in cereals is highly variable in wholegrain and bran, and strongly dependent on cereal variety, milling procedure and processing conditions. Dietary fiber in the grain is concentrated in the bran representing between 18.1 and 86.7% of the weight.[18] The insoluble dietary fiber content varies with the type of cereal from 1.9 g 100 g^{-1} in rice, to approximatively 15 g 100 g^{-1} in rye. Wheat, maize and oat have intermediate content with values of 10.8, 8.5, and 5.8 g 100 g^{-1}, respectively. Rye not only has higher fiber content but is also rich in beneficial compounds such as folate, phenolic acids, alkylresorcinols, lignans, phytosterols, tocopherols, tocotrienols, and inositol hexaphosphate.[19] This fact confers one of the paradoxical effects for risk/benefit evaluation of the mitigation strategies of acrylamide formation in cereals.

3.3 Occurrence, Exposure and Regulation

The food industry, academia, research centers, national food safety agencies, administration and supranational organizations (*i.e. Codex Alimentarius* FAO-WHO) have undertaken extensive efforts to promote investigations to reduce acrylamide levels in processed foods. Particularly, the industrial sector has had a proactive position, and particularly for food commodities based on potato, coffee and cereals, as well as those intended for infant consumption. In 2005, Food Drink Europe (formerly, CIAA) released the first edition of the acrylamide toolbox. The objective of this guide was to provide practical tools that the food industry could implement to reduce acrylamide formation. However, the extent of acrylamide mitigation strategies may be constrained by budgetary measures. The food industry in general works to very small profit margins per item produced. A large food producer may be able to implement expensive mitigation strategies in a relatively short time-frame. However, small artisan producers may find that implementation of acrylamide mitigation strategies impact the price of their goods. For cereal-based foods, the primary efforts are focused on wholegrain and bran products. Other strategies are focused on recipe changes and modifications in the product design or in the processing conditions. *The Acrylamide Toolbox*[20] is updated in a timely manner according to advances in the state-of-the-art processing methods. The latest version is the 13th, dated January 2014, although a further release is in preparation.

The investigations carried out by stakeholders covering the identification of formation pathways, intermediates and precursors, factors controlling the reaction, strategies of mitigation, techniques of analysis, epidemiology, exposure to population groups, and risk evaluation have been substantial. This knowledge was compiled by the European Food Safety Authority (EFSA) Panel on Contaminants in the Food Chain (CONTAM) that adopted a scientific opinion on acrylamide in food in June 2015.[21] EFSA evaluated results of acrylamide content from thousands of food commodities marketed in EU countries. In that survey, the cereal foods represent 11.48% of the total dataset of food categories used for the calculation of chronic dietary acrylamide exposure within the population. The subgroup of biscuits, crackers, crispbread and similar foods represent 41.4% of the samples collected in the cereal group, followed by breakfast cereals with 24.7% of market share. The sampling covered a period from 2010 to 2013 and was performed by the official monitoring programs and surveys of European countries, as well as by food associations (European Coffee Federation, European Breakfast Cereal Association, European Snack Association, Food Drink Europe), and two national associations (Finish Food and Drink Industries' Federation, and the German Plant Bakeries Associations). In addition, toxicological studies performed in experimental animals (carcinogenicity and genotoxicity), toxicokinetic trials (absorption, distribution, metabolism, and excretion), biomarkers of exposure, and epidemiological studies have determined a complete risk characterization. However, although the epidemiological associations have not definitively demonstrated acrylamide to be carcinogenic to humans, the margins of exposure (MOEs) indicate a high concern for neoplastic effects based on animal evidence.[21]

In 2007, the European Commission adopted recommendations on the level of acrylamide in foods, establishing indicative values for the main food sources of exposure to this contaminant. These indicative values were revised in the years thereafter. Member States of the European Union were recommended to carry out additional surveillance control in cases where the levels of acrylamide in a foodstuff exceeded acrylamide indicative values. In 2017, the European Commission enacted Acrylamide Regulation 2017/2158 that compels food processors and food business operators in Europe to reduce the presence of acrylamide according to the ALARA principle (as low as reasonably achievable). The Regulation includes measures proportionate to the size and nature of the establishments without compromising the taste, texture and appearance of the products.[22] The Regulation established mitigation measures for the reduction of acrylamide in foods as well as new benchmark levels for a number of food categories. The levels will be revised after three years. The regulation for acrylamide in the EU came into force on 11 April 2018.

Cereal foods are distributed into different groups, as described in Table 3.2. A benchmark level for acrylamide of 50 μg kg^{-1} was established for soft bread. The definition of soft bread includes wheat-based bread where the cereal ingredient is at least 50% wheat. In cases where the main grain is not wheat, the benchmark level is set at 100 μg kg^{-1}. In the first announcement of indicative values for soft bread, the level was set at 150

Table 3.2 Distribution of the different groups of cereal-based foods in the dataset and the mean acrylamide content (μg kg^{-1}) used by the EFSA CONTAM panel for the estimation of acrylamide exposure of the population. Number of samples (*n*), sum of samples from European countries and food associations (*N*). Data from ref. 21.

Food group	European countries (*n*)	Food associations (*n*)	*data* (*N*)	Relative to cereals (%)	Relative to all foods (%)	Acrylamide mean (μg kg^{-1})
Soft bread	535	8	543	10.89	1.25	42
Breakfast cereals	561	669	1230	24.67	2.83	161
Biscuits, crackers, crispbread	1974	91	2065	41.42	4.76	265
Coffee substitute based on cereals	20	0	20	0.40	0.05	510
Cake and pastry	198	0	198	3.97	0.46	66
Processed cereal-based baby foods	394	392	786	15.77	1.81	73
Other products based on cereals	0	143	143	2.87	0.33	68
Subtotal	3682	1303	4985	100.00	11.48	

μg kg^{-1}, and in the intervening 10 years later this value has been significantly decreased. According to EFSA (2015),[21] the average acrylamide content in wheat soft bread was 38 μg kg^{-1} ($n = 302$), close to the benchmark levels for this product. Lower levels were observed in other soft bread (mainly made from rye), where the mean value reported was 57 μg kg^{-1} ($n = 107$). Breakfast cereals are another category of cereal food under regulation. Breakfast cereals include toasted flakes and gun-puffed breakfast cereals. The cereal present in the largest quantity, as well as the presence of wholegrain, determine the subcategory. The mean acrylamide content in maize, oat, spelt, barley and rice based breakfast cereals is 102 μg kg^{-1}, 211 μg kg^{-1} for bran products and whole grains, and 170 μg kg^{-1} for wheat and rye based breakfast cereals (Table 3.2). Again, these values are nearly 50% lower than the benchmark levels. In the case of biscuits, wafers, crackers, crispbread and related food categories, the situation is more complex because of the presence of many other ingredients in the recipes. EFSA estimates a mean acrylamide content of 201 μg kg^{-1} for biscuits and wafers, 231 μg kg^{-1} for crackers (not potato based), 171 μg kg^{-1} for crispbread, and 407 μg kg^{-1} for gingerbread.[21] These values are significantly lower than the benchmark value established in the regulation for acrylamide (Table 3.3).

Cereals belong to the *Gramineae* family (*Poaceae*) and are commonly known as grasses. Cereals such as wheat, barley, oat, rye, and maize are staple foods for the population of European countries. Rice, wheat, and maize provide 93% of the total cereal calories for Asian, European and American populations, respectively.[23] Global cereal consumption provides approximately 45%

Table 3.3 Indicative values for acrylamide (μg kg^{-1}) in cereal-based foods according to the recommendations from 2011 and 2013, and benchmark levels for the presence of acrylamide in foodstuffs according to the regulation from 2017.

Cereal-based foods	2011	2013	2017
Soft bread			
(a) Wheat-based bread	150	80	50
(b) Soft bread other than wheat-based bread	150	150	100
Breakfast cereals (excluding porridge)			
- bran products and whole grain cereals, gun puffed grain	400	400	300
- wheat and rye based products	400	300	300
- maize, oat, spelt, barley and rice based products	400	200	150
Biscuits and wafers	500	500	350
Crackers with the exception of potato-based crackers	500	500	400
Crispbread	500	450	350
Ginger bread	—	1000	800
Products similar to the other products in this category	500	500	300
Coffee substitutes			
(a) Coffee substitutes exclusively from cereals	—	2000	500
(b) Coffee substitutes from a mixture of cereals and chicory	—	—	[a]
(c) Coffee substitutes exclusively from chicory	—	4000	4000
Baby foods*, processed cereals-based foods for infants and young children excluding biscuits and rusks**	80* 100**	50	40
Biscuits and rusks for infants and young children	250	200	150

[a]Depends on the relative proportion of the ingredients in the final product.

of protein and energy necessary for the human diet and only approximately 7% of the total fat. The specific contribution of wheat to daily food intake corresponds to approximately 20% of the required energy and protein for the human diet.[24] However, the contribution of cereal-based foods to the total dietary exposure to acrylamide varies greatly among European countries, and within the different types of cereal groups. In 2001, EFSA estimated that the total contribution of cereal products to dietary acrylamide intake for adults (18–64 years of age) in France, Germany, Sweden and the UK was 40.9%, 45.4%, 41.2% and 31.9%, respectively.[25] In the category of cereal products, biscuits, crispbread, bread, breakfast cereals and muesli were included as relevant food groups. However, the contribution within cereal food groups was significantly higher for bread in France and Germany, for breakfast cereals in the UK, and for muesli in Sweden. In 2015, EFSA recalculated the contribution of each food group to the total exposure to acrylamide for each population group. This contribution was calculated as the ratio between the average acrylamide intake resulting from the consumption of the food group and the total average exposure to acrylamide. The exposure to acrylamide was assessed by multiplying the mean daily consumption of each food by the corresponding mean occurrence level, and dividing the results by the body weight.[21] There are some clear conclusions: (a) the group of processed cereal-based baby foods represents 30% of the total acrylamide exposure for infants, (b) for the age groups of adults, elderly and very-elderly, the relative contribution for soft bread, biscuits (including crackers, crispbread, *etc.*), breakfast cereals, and coffee substitutes, represents 9.9–23%, 2.4–24%, 0.3–10%, and 0–8.7%, respectively.

3.4 Mitigation: Organoleptic and Nutritional Consequences

There are three key parameters to be considered before the introduction of a mitigation strategy for acrylamide: (i) the reduction rate obtained, (ii) potential side effects, and (iii) the applicability, scaling-up and economic impact of its implementation.[26] Owing to the fact that cereal foods include products with different ingredients, recipes and processes, there is no single solution to mitigate acrylamide. Within the acrylamide toolbox and the Commission Regulation (EU) 2017/2158, reduction of acrylamide in cereal-based products could be achieved through intervention at various stages of production by modifying agronomic factors, recipe and product design, processing and final preparation strategies (Figure 3.2).[20,22]

3.4.1 Agronomical Factors

As previously outlined, the reduction of asparagine content in the flour is the primary strategy to prevent the formation of acrylamide during the processing of cereal-based foods. The free asparagine content in cereal crops

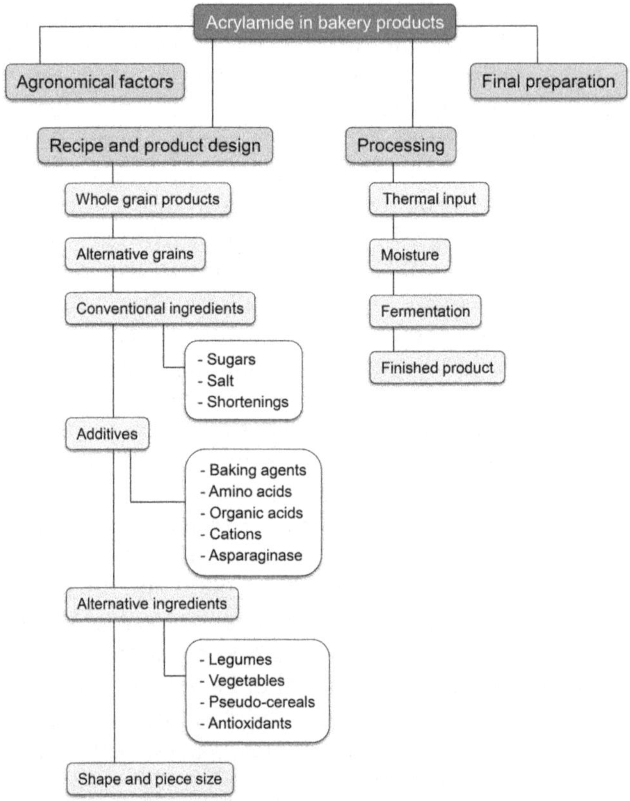

Figure 3.2 Interventions for acrylamide mitigation in bakery products.

is not easy to control because of several sources of variation, such as the type of grain, variety, growing conditions, and climate. In general, acrylamide may be reduced by selecting suitable cereal varieties with low asparagine content.[4] Weather conditions in the harvest year, including growth temperature and humidity, affect protein and amino acid contents. This fact, in turn, may determine sprouting in cereal production, which doubles protease activity resulting in high asparagine levels. Climatic conditions are more difficult to control but appropriate fertilization could easily be selected. Nitrogen fertilization, use of which is recommended to enhance crop yield and flour quality, has a strong impact on crude protein and free asparagine content in wheat flours, with up to a four-fold increase when 0 kg N ha^{-1} was compared with 220 kg N ha^{-1}.[10] In contrast, sulfur fertilization may reduce asparagine content in flour. Muttucumaru *et al.*[27] have described reductions from 5286 to 723 μg kg^{-1} in wheat flour heated at 160 °C for 20 min, when wheat was treated with 40 kg S ha^{-1}. However, the same treatment in rye did not affect the final levels of acrylamide in heated flour. This can be explained because the treatment and also the genotype and the interaction between them have a strong effect on the characteristics of the flour.[12] The cultivation of cereal through organic farming or conventional agriculture may affect the asparagine content in the flour,[28] but the information reported is still controversial. Differences are related to the application of best practice for crop protection measures to prevent fungal mycotoxin infection, since free asparagine may increase in response to pathogenic attack infections.[29]

3.4.2 Recipe and Product Design

3.4.2.1 *Wholegrain Products*

Since asparagine content is higher in the outer layer of the grain (Table 3.1), limiting the use of wholegrain and/or bran and germ and increasing the proportion of the endosperm part in the flour, will decrease the amino acid content in the formulation, resulting in controlled levels of acrylamide in bakery products. The acrylamide regulation describes how products made up of wholegrain cereals and/or with high levels of cereal bran have greater levels of acrylamide.[22] Acrylamide content in crispbread prepared from rye flours with extraction rates of 70, 85, 95, and 100% decreased directly with the decrease in the extraction rate.[30] Flour with an extraction rate of 100% had almost twice as much free asparagine as a flour with a 70% extraction rate. On the other hand, during milling, starch in wheat grains is subjected to compressive and shear forces, which results in different degrees of damage to starch granules. Springer *et al.*[31] claimed that baked products made from less intensively milled flour had a reduced content of acrylamide. Wang *et al.*[32] also observed higher levels of acrylamide in bread crust and crumb with increased content of damaged starch in the wheat flour.

Both the fractionation efficiency of the milling and the recipe directly affect acrylamide levels in bakery products. In parallel, the organoleptic and nutritional profiles may be significantly compromised. Consumption of wholegrain cereals and bran are one of the most beneficial important sources of fiber in the diet. A daily intake of dietary fiber ranging from 25 to 38 g per day is recommended to promote intestinal health and for prevention of Type 2 diabetes, colorectal cancer, cardiovascular disease and obesity.[33] The milling process removes the bran and germ of the grain, which are rich sources of dietary fiber, fructans, and other health-promoting bioactive compounds. For that reason, unlike refined grains, wholegrain flours are known to improve glycemic control and insulin sensitivity.[34] The beneficial effects attributed to cereal dietary fiber are due not only to the polysaccharide moiety, but also to the phenolic compounds linked to the fiber core. The dietary fiber-phenolic compounds (soluble and insoluble) may exert *in vivo* antioxidant protection acting at the intestinal environment level. Free phenolics, which are released from the soluble fiber by the action of intestinal and microbial esterases, can be absorbed through the intestinal wall and passed into the bloodstream, where they exert their health benefits at a systemic level.[35,36] Since the current consumption of dietary fiber in the average population is about 18 g per day,[33] from a nutritional perspective, it does not seem appropriate to decrease the presence of fiber-enriched ingredients in cereal-derived products despite this being a mitigation strategy to control acrylamide formation.

3.4.2.2 *Replacing Wheat with Alternative Grain*

In multicereal formulations, the total or partial replacement of cereal flours rich in asparagine, such as rye and wheat, with an alternative grain could be considered in order to mitigate acrylamide formation. Cheng *et al.*[37] proposed the following index for acrylamide formation from snack foods made of common grains: rice flour < corn flour < wheat flour < wheat/potato mixed flour < rice/potato mixed flour. The case of rye deserves special attention. Rye is widely distributed in the diet of North and East European countries.[38] Rye proteins have a relatively high lysine content compared with wheat, resulting in a higher nutritional quality. Moreover, rye provides the cereal-derived products whose sensorial attributes consumers appreciate, such as texture, flavour, and colour.[39] However, substitution of wheat by rye flour in the recipe for bread making has led to higher values of acrylamide and an emerging processing contaminant, hydroxymethylfurfural (HMF).[40] On the other hand, the replacement of wheat with rice flour may be an effective measure if acceptable from a quality perspective, but additional concerns must be taken into account.[20] In particular, when this reformulation is performed in infant biscuit products the potentially higher arsenic level in rice compared with wheat should be considered.[41] In any case, the choice of the grain defines the final product and it is not possible to simply replace one grain by another without sensory consequences that impact on product identity and acceptability to consumers.

3.4.2.3 *Conventional Ingredients*

Dough mixtures usually contain ingredients other than flour and water that also influence the content of acrylamide in the finished product. These conventional ingredients are sugar, salt, and shortening agents.

Sugars present different relative reactivity with regard to acrylamide generation and, among them, fructose shows the highest reactivity. The addition of 1% fructose to the flour in the formulation of soft biscuits increased acrylamide formation by about 20%.[42] The replacement of inverted sugar syrup with a sucrose solution in the dough has led to an acrylamide level of 46 $\mu g\ kg^{-1}$, which was 70% lower than that found in the standard formulation.[43] Similarly, Vass *et al.*[44] and Gökmen *et al.*[45] reported that the replacement of reducing sugars by sucrose reduced acrylamide content by 60% in crackers and 50% in cookies. These findings suggest that replacing inverted sugar syrup in the dough recipe is an effective strategy to significantly reduce the acrylamide content in bakery products. However, in some cases, sucrose may be hydrolyzed to form reducing sugars during thermal treatment, showing similar reactivity to glucose and fructose. This fact has been corroborated by Taeymans *et al.*[11] in a washed whole-wheat flour matrix and by Nguyen *et al.*[46] in biscuits. Other factors to take into account is the control of sugars added into the formulation. In this respect, contradictory results have been reported. Gökmen *et al.*[45] indicated that increasing the amount of sucrose from 10 to 35 g in the recipe for cookies almost doubled the amount of acrylamide formed during baking. In contrast, Surdyk *et al.*[47] observed that addition of different amounts of fructose to yeast-leavened wheat bread did not influence the content of acrylamide.

Salt (sodium chloride; NaCl) is an essential ingredient during dough formation since it induces important changes in rheological, technological, and sensory properties in the final product. Salt is not only related to seasoning or used as a flavour enhancer, it is also involved in the formation of the gluten matrix, increasing stability, flexibility and resistance of the dough as well as decreasing stickiness and water absorption.[48] Salt has a considerable inhibitory effect on acrylamide formation based on its capability to reduce amylase activity and in turn decrease the formation of reducing sugars,[7] as well as in its ability to accelerate the elimination of acrylamide *via* polymerization within the dough matrix.[49] However, it is important to control the amount of salt in the dough during the yeast fermentation step and also the conditions of baking. The addition of 1–2% of sodium chloride to a bread mix reduces acrylamide levels in baked rolls while higher salt concentrations resulted in increased acrylamide formation as a consequence of the inhibition of the yeast growth that reduces the metabolization of asparagine.[7] The effect of NaCl on acrylamide formation also takes place during baking. Van Der Fels-Klerx *et al.*[50] reported lower acrylamide concentrations in biscuit formulations with NaCl, but inhibition was only exhibited when baked at 180 and 190 °C, but not at a higher temperature of 200 °C.

Despite increasing the salt content in the dough being adopted as a successful acrylamide mitigation strategy, a major drawback of this approach is the increase of sodium in the final product and thus the salt consumption by the population. Sodium intake is of great concern to the World Health Organization (WHO). In fact, WHO recommends a sodium consumption of less than 2 g per day to exert a decrease in blood pressure and risk of cardiovascular disease, stroke and coronary heart disease in adults.[51] Current reports point to a global sodium intake in the general population of 3.95 g per day (95% uncertainty interval: 3.89–4.01 g per day),[52] nearly twice the WHO recommendation. The value is higher in developed countries due to the consumption of highly processed foods. Some investigations support the notion that nowadays grain-based products, such as bread and breakfast cereals contribute an increase of 25% salt consumption.[53] Therefore, from a health perspective, strategies to control acrylamide formation by increasing sodium salts in the formulation should be undertaken with a degree of caution.

Shortenings enhance the sensorial and rheological properties of confectionery and bakery products. However, they may affect the acrylamide produced in products depending on the type and the composition of vegetable oil or fat used. Capuano *et al.*[54] showed that acrylamide levels were higher in model systems incorporating sunflower oil than in those containing palm oil. Arribas-Lorenzo *et al.*[55] evaluated the effect of adding virgin olive oils with different phenolic content in the formulation of cookies and observed that formulations with high phenolic compounds content resulted in acrylamide levels 20% lower than those with a low content of these compounds. These authors also investigated the effect of oil oxidation level on acrylamide formation in cookies and claimed that lipid oxidation products may lead to a significantly high increase in the formation of acrylamide during baking. This statement can be explained as carbonyl compounds derived from lipid oxidation during food processing can degrade asparagine to acrylamide and then lipids can contribute to acrylamide formation during heat treatment of fat-rich foods.[56] The effect of oil oxidation levels on acrylamide formation can be even more evident if sugar is removed from the formulation.[54] The use of unsaturated vegetable oils rich in phenolics will impair acrylamide formation but will result in adverse sensorial attributes and a concomitant decrease in the shelf-life of the product.

3.4.2.4 Additives

The incorporation of baking agents or leavening agents such as biological (fresh or dehydrated yeast) or chemical (*i.e.* baking soda) leavening causes expansion of the dough. The production of gas from yeast or baking soda promotes an increase in volume and produces a desirable shape and crumb texture.[57] However, these additives strongly influence the acrylamide formation in the bakery product. The activity of the yeast (*Saccharomyces cerevisiae*) in sourdough for bread-making reduces the initial asparagine content of the flour. Chemical leavening is commonly used for biscuits due to a faster, more convenient and often more consistent method of leavening. The use of ammonium bicarbonate (NH_4HCO_3) has been demonstrated to promote the formation of acrylamide in cereal products.[58,59] This could be attributed to an activation of reducing sugars by ammonia or the formation of dicarbonyl compounds (glyoxal and methylglyoxal), which are potent precursors of acrylamide in the presence of asparagine.[60] Replacing ammonium bicarbonate by sodium hydrogen carbonate greatly reduces the acrylamide formation by raising the pH of the system and, at the same time, promotes acrylamide elimination.[61] Reductions of around 60% in gingerbread[62] and cookies[58] with sodium bicarbonate addition have been reported. Despite these positive aspects, combinations of both baking agents are often required in order to achieve the correct balance of gas release during baking and optimum organoleptic characteristics. Reformulated products should be carefully considered due to changes in flavour, colour, aroma and texture that may lead to possible rejection by the

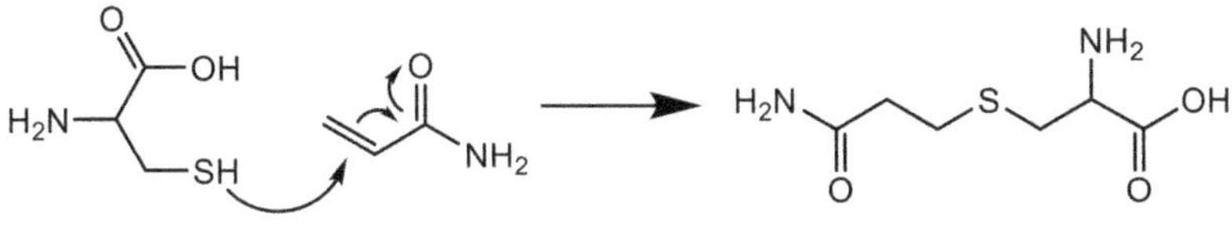

Figure 3.3 Mechanism of Michael addition between cysteine and acrylamide.

consumer. Kukurová *et al.*[58] stated that $NaHCO_3$ addition to bakery products leads to more pleasant sensory attributes. In contrast, biscuits with high doses of sodium bicarbonate have been reported to have an alkaline taste, a yellowish crumb, and surface coloration and an unpleasant taste known as soda bite.[63] On the other hand, the employment of sodium bicarbonate must be handled with care since it increases the sodium content in the formulation, with organoleptic changes and nutritional consequences as described previously.

Adding certain amino acids to the formulation of bakery products has been demonstrated to decrease acrylamide content. Amino acids may compete with asparagine in the Maillard reaction, avoiding acrylamide formation; furthermore, they may react with acrylamide through Michael addition after its formation (Figure 3.3), promoting the elimination of acrylamide. Among them, glycine has been reported to be effective in acrylamide mitigation. Capuano *et al.*[64] observed that the addition of 5 g of glycine per kg of flour to the basic formulation of crispbread significantly decreased levels of acrylamide by up to 60% after different toasting times. Similar results were observed by Amrein *et al.*[62] in gingerbread when 10 g of glycine was incorporated per kg of dough. In this study, acrylamide content was reduced to one-third and, in contrast, acrylamide content was not lowered by the addition of L-glutamine, L-lysine and glycine at concentrations of 2 g kg^{-1} of dough. In agreement with these findings, Brathen *et al.*[65] described acrylamide reduction by up to 80% and 90% when glycine was applied to the dough prior to the fermentation and production of flat breads and breads, respectively. However this modification may have negative effects on the sensory properties of the products, since the reaction between glycine and reducing sugars increased levels of odour-active alkylpyrazines.[66] Several experiments with the addition of L-cysteine and methionine have shown positive effects in the reduction of acrylamide in gingerbread[62] and crackers[61] but no effect was observed when cysteine was added directly to the dough of crispbread.[67]

The addition of organic acids in bakery products can minimize acrylamide content through the reduction of acrylamide formation rates but also by enhancing the degradation of acrylamide already formed. This was demonstrated by Graf *et al.*[43] adding tartaric acid to the dough of bread. Results, however, could be affected by a partial replacement of ammonium bicarbonate with sodium bicarbonate. Amrein *et al.*[62] observed that the presence of citric acid in gingerbread decreased the pH in the system and subsequently lowered acrylamide formation. On the other hand, ascorbic acid added to cracker models led to increased elimination and decreased formation, thus reducing the overall content of acrylamide in the product.[61] A negative effect associated with the addition of acids is a modification of the sensorial properties and, therefore, the amount to be added should be limited. In addition, adding organic acids in certain bakery products, in combination with fermentation, can lead to an increase in 3- monochloropropanediol, another process contaminant.[68]

Divalent cations have been known to efficiently lower acrylamide concentration in bakery products. Addition of calcium (Ca^{2+}) has been reported to decrease the level of acrylamide by up to 30% in bread and crackers.[69] This is based on the theory that divalent metal ions added to the food product may improve high-temperature stability to the asparagine/matrix interactions, decreasing the opportunity for reactions with carbonyl precursors to generate acrylamide.[70] However, the drop in the pH value in the system should also be considered. Among inorganic salts, the most effective acrylamide inhibitor in cereal matrices has been suggested to be calcium chloride ($CaCl_2$). In this regard, Mesías *et al.*[71] observed that acrylamide content was reduced by up to 58% in cookies containing $CaCl_2$ instead of NaCl. However, cookies exhibited a dark colour and unpleasant organoleptic properties as well as a significant increase in the content of 5-HMF. Reductions of 60 and 36% in acrylamide formation have also been described when $CaCl_2$ was added to dough of bakery products and in cooked wheat products, respectively.[68,72] Although with less effectiveness, monovalent cations have a considerable inhibitory effect on the formation of the contaminant, as mentioned previously (Section 3.4.2.3).

Nowadays, the most effective mitigation strategy for acrylamide reduction in cereal-based foods is the pretreatment of the dough with the enzyme asparaginase.[58,62,73–75] Asparagine is hydrolyzed into aspartic acid by the action of the enzyme, thus inhibiting acrylamide formation. The asparaginase activity will depend on the enzyme concentration, time and temperature of enzyme incubation before thermal processing of food and the pH at which the reaction occurs. Water activity of the food matrix also greatly affects asparaginase efficacy, since it impacts the mobility of the substrate and enzyme.[75] Asparaginase is especially effective in bakery products and biscuits but its use is limited in breakfast cereals due to the moisture level. Anese *et al.*[75] evaluated the effect of adding asparaginase in the range of 100–900 U kg^{-1} of flour in short dough biscuits, together with incubation times and temperatures between 10 and 30 min and 20–37 °C, respectively. In that study, 500 U kg^{-1} of asparaginase with 10 min incubation time at 20 °C resulted in the minimum acrylamide concentration in biscuits. Kukurová *et al.*[58] found that the highest efficiency of acrylamide reduction in cookies was achieved after heating times of 11 and 15 min, with 60 min asparagine pre-treatment at pH 6.78. The prolongation of the asparaginase incubation time intensified the conversion of asparagine. However, although prolonging the incubation time resulted in a depletion of final acrylamide content at all pH values, the efficiency of the enzyme activity differed with pH level. This strategy should be taken into account when additives for cereal-based products are selected. In this sense, cookie dough supplemented with ammonium bicarbonate (NH_4HCO_3) and sodium bicarbonate ($NaHCO_3$) resulted in samples with a lower enzyme activity of asparaginase due to higher pH values.[76] The enzyme treatment at low dosage rates had no significant impact on final organoleptic properties of biscuits and cookies.[58,62]

In recent years, new sources of asparaginase for acrylamide reduction apart from *Aspergillus orizae* and *Aspergillus niger* have been identified. Thereby, reductions of 97% in bread crust and 73% in bread crumb were achieved by using 300 U of asparaginase produced from *Clostridium* sp.[77] Huang *et al.*[78] extracted asparaginase from *Rhizomucor miehei* and demonstrated that a concentration of 0.5 U g^{-1} of flour mitigated acrylamide in bread (40%) and biscuits (15%). If the enzyme was used at higher concentrations (10 U g^{-1} of flour), 80% inhibition was achieved in both products. Amrein *et al.*[62] reported a 55% acrylamide reduction in gingerbread by the addition of asparaginase from *Escherichia coli* (4 U kg^{-1}), which led to the hydrolysis of approximately 75% of the free asparagine.

3.4.2.5 *Alternative Ingredients*

The cereal industry is constantly evolving in order to offer healthier alternatives that provide consumers an improved nutritive quality of products. For this reason, some cereal-based products have begun to include in their formulation legumes, pseudo-cereals and vegetable extracts rich in antioxidants in order to complement their nutritional features, especially in terms of amino acid profile.

Cereals provide a good level of proteins to the daily diet, but are particularly deficient in some essential amino acids, such as lysine and threonine. Additionally, some formulations aim to incorporate other cereal flours, such as rice, maize, barley, sorghum, and rye, producing bread, biscuits, and other bakery products. In the last decade, legume flours have been considered for fortification of wheat foods, usually at 5–25%, since legume proteins are rich in lysine and tryptophan. Legumes that can be used are soy (*Glycine max*), bean (*Phaseolus vulgaris* L.), broad bean (*Vicia faba* L.), chickpea (*Cicer arietinum* L.), lentil (*Lens culinaris*), lupine (*Lupinus* sp.), pea (*Pisum sativum* L.), vetch (*Vicia sativa* L.), and peanut (*Arachis hypogaea*). However, the free asparagine content in some legumes, such as lentil, may reach up to 3500 mg kg^{-1}.[79]

Pulses have an appreciable quantity of all the essential amino acids excluding sulfur-containing ones, which can be balanced by combination with cereals in the normal daily intake.[80] The effect of legume inclusion on acrylamide formation depends on the specific type of seed. Galani *et al.*[81] studied the acrylamide forming potential of cereals, legumes and roots and tubers and established that the combination of pulses (chickpea and

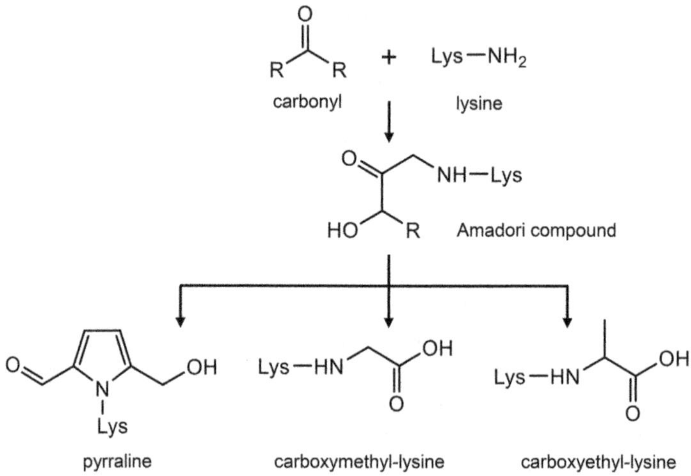

Figure 3.4 Selected structures of protein cross-linked lysine derived AGEs formed through the Maillard reaction from carbonyl compounds.

soybean) and minor cereals (millet and sorghum) in cereal-based foods not only improves their nutritional value but also reduces acrylamide formation. Dried soybean based snacks have been shown to have low acrylamide levels ranging from <5–98 μg kg^{-1}.[37] However, although this seed has important amounts of lysine, the concomitant formation of other food processing contaminants such as Amadori compounds or advanced glycation end-products (AGEs) such as carboxymethylysine is a little known aspect that should be considered (Figure 3.4). In fact, the recent study by Milkovska-Stamenova *et al.*[82] has reported 147 glycation sites in 26 soy proteins and 112 AGE-derived modification sites in 41 soy proteins of natural and differently sweetened soy milk products including soy-based infant formulas. Other authors have confirmed that chickpea flour and protein may be a new way to lessen acrylamide production in bread and snacks, but they also underline that the sensory and textural properties of the resulting products are not always liked by consumers.[83]

The inclusion of germinated seeds of lentils in recipes does not seem an appropriate strategy to control acrylamide occurrence because asparagine is quantitatively by far the most important protein amino acid, with concentrations ranging from 19 to 62 mg g^{-1}.[84] The case of pea seems to be more promising since it has been reported to have the highest natural asparaginase activity of the stated plant sources.[85] Some investigations have shown that the reduction of acrylamide levels in white bread with added germinated pea flour was not significant and that the presence of this legume decreased consumer acceptance. However, it was found that the acrylamide level can be reduced by 57 and 68% with addition of 5% pea flour in wheat bran and wholegrain breads, respectively, without any negative impact on colour and sensory properties.[74] All the breads in the study mentioned were made with potassium concentrations above 20 mM, since this has been shown to maximize plant asparaginase activity.[85] Therefore, a significant reduction in acrylamide levels is associated with major potassium participation in the dough, which arises from the high mineral content of the outer layers of the grain in the case of wheat bran and wholegrain breads. Some other legume seeds such as *Vicia faba* (faba bean) have also been reported to have significant amounts of asparaginase,[86] so that its potential as an ingredient in order to control acrylamide formation in alternative cereal-based formulations should be considered. In any case, it is important to keep in mind that an asparaginase strategy is only efficient when the cereal is in a flour dough format and the process enables sufficient time, temperature and moisture content to promote the enzyme activity.[20]

Lastly, from a food security point of view, the introduction of legumes in recipes, besides reducing acrylamide in the novel product, must not lead to the occurrence of any other food processing contaminant. Information with regard to this problem is scarce in the scientific literature. An investigation has studied the effects of the inclusion of seven different legumes (black soybean, chickpea, green pea, lentil, mung bean, yellow pea, and soybean) in an asparagine–glucose model system. When the defatted-dehulled flours of the legumes were present in the mixture, acrylamide production was further inhibited in some treatments but this occurred along with an increase in the content of the processing contaminant HMF.[87] The recent study by Pedreschi *et al.*[88] has provided evidence of the mitigation ability of tara pod, from the leguminous tree *Caesalpinia spinose*, on acrylamide and HMF formation in Chilean bread, while preserving its sensory attributes. However, it must be mentioned that only the polyphenolic extract of this pod was used and not the complete seed.

Other alternative ingredients introduced in the formulation of some cereal-derived products are vegetables, especially in the case of the snack industry. A new trend is the elaboration of innovative salty snacks using cereal flours combined with roots or vegetables (veggie chips) in order to replace potato or some grains that have high acrylamide-forming potential. In this regard, our research team has established that pumpkin and carrot are less prone to the formation of the contaminant than potato during a standardized frying process.[89]

Recently, pseudo-cereals such as chia, teff, amaranth or quinoa have been introduced in the formulations of innovative cereal-based products. Due to their nutritional value and chemical composition, different medicinal properties have been attributed to pseudo-cereals and they have been considered as new functional ingredients.[90,91] Therefore, their incorporation in the formulation of certain foods may be particularly desirable from a nutritional and health point of view. Similarly, the addition of whole pseudo-cereals seeds such as buckwheat, amaranth, teff and quinoa, which are naturally gluten-free, is increasing as an alternative mitigation strategy for coeliac disease.[79] How these supplementations affect acrylamide or other thermal processing contaminants is still quite unknown. In this respect, Mesías *et al.*[92] demonstrated that, although antioxidant capacity, phenolic compounds, protein, fiber and polyunsaturated fatty acid content were improved in biscuits by the incorporation of chia flour, the new formulation increased the acrylamide content. Other contaminants such as HMF and furfural were also increased, as well as some precursors (dicarbonyl compounds such as methylglyoxal and glyoxal). Moreover, lipid oxidation products, particularly polymerization compounds, were accelerated in chia-enriched biscuits, decreasing the shelf-life of the product by promoting a rapid rancidity under accelerated storage conditions.[92] The inclusion of teff in the formulation could be a corrective strategy to keep acrylamide levels controlled, since investigations performed in our research group have shown a low presence of the compound in ready-to-eat cereals made of teff as a unique ingredient (26 μg kg^{-1}; unpublished results). However, from a risk/benefit perspective, previous work has stated that replacement of wheat flour with teff in a heated model system in the presence of glucose induced higher HMF and furfural formation, as well as the AGE precursor 3-deoxyglucosone.[93] Amaranth protein isolate has been successfully included in the recipe of cookies, fried tortilla chips, and baked tortilla chips to decrease acrylamide production with a decrease ranging from 89–51%. Cookies maintained proper colour and texture, but hardness of the tortilla chips was significantly raised.[94] The quinoa seed (*Chenopodium quinoa*) is high in lysine, methionine, and cysteine, making it complementary to both other grains and legumes. However, it is also important to evaluate the level of acrylamide precursors in the dough since proteins improve the amino acid balance but, at the same time, higher levels of acrylamide will be generated from the same process. In this sense, quinoa lipids have a high degree of unsaturation[95] that potentially will increase acrylamide formation during baking, arising from lipid oxidation compounds, if non-defatted quinoa flour is used. Kamut is not a pseudo-cereal but one of the most ancient varieties of wheat[96] that has begun to be included in innovative recipes for cereal-derived formulations. Despite the nutritional attractiveness of this ingredient, a high acrylamide presence was detected in kamut breakfast cereals analyzed in a prospective study developed by our research team (639 μg kg^{-1}; unpublished results), however this data must be further investigated.

Acrylamide formation may be inhibited by the addition of antioxidants. These compounds can react with acrylamide precursors or intermediates, which, in turn, may inhibit the overall rate of the Maillard reaction. Significant acrylamide reduction has been described in fried bread sticks supplemented with antioxidants from bamboo leaves and green tea[97] and in bread when antioxidants from rosemary extract, rosemary oil and dried rosemary leaves were added to wheat dough.[98] Similar results were reported by Levine and Smith[61] in a cracker model system, where the addition of ferulic acid reduced net acrylamide formed. Controversial results have been reported by Açar and Gökmen[99] and Capuano *et al.*,[40] who did not observe acrylamide inhibition in bread if wheat flour was enriched with grape seed or green tea extracts, respectively.

3.4.2.6 Shape and Piece Size

Controlling the dimensions of the bakery product is another strategy to monitor acrylamide formation. In this regard, bread volume measurements in addition to time and temperature of baking have been demonstrated to have a significant effect on acrylamide levels in the baked product.[8] In addition, the product thickness also determines the rate of formation of acrylamide, which is linked to the rate of drying. It has been reported that dough baked at 180 °C for 8 min having a thickness of 1 mm reaches an acrylamide content of 411 µg kg^{-1}. However, dough having thickness of 10 mm baked for 15 min under the same conditions does not generate acrylamide.

3.4.3 Processing

3.4.3.1 Thermal Input and Final Product Moisture

Acrylamide formation is directly linked to time–temperature conditions during baking so that increasing baking time or temperature significantly enhances acrylamide formation in the baked product. At a constant temperature, acrylamide levels increase with the duration of the thermal treatment and, at the same time, higher temperatures promote the higher formation of acrylamide.[100] Maximum acrylamide formation is reached at high temperatures during a short period of time and then, levels decrease due to a high rate of degradation. For example, gingerbread baked at 180 °C and 200 °C for 20 min and bread baked at 260 °C for 15 min displayed the highest levels of acrylamide, decreasing with prolonged times.[62,101] In general, lower acrylamide levels are found in cereal products baked at lower temperatures for a longer time. Nevertheless, these conditions should be critically evaluated due to associated changes in flavour and colour in insufficiently browned products. In the case of bread, most acrylamide is found in the outer crust layer, while crumb only contains trace amounts. This observation is linked to process temperatures (lower temperatures are ascribed to inner parts of the product as compared with outer parts) but also to the moisture of the product since high moisture content inhibits acrylamide formation. Elmore *et al.*[13] reported relatively low values for acrylamide in cakes baked in an oven at 180 °C for 5–60 min when moisture content was above 5%, whereas levels increased when the moisture content fell below 5%. In a similar way, Taeymans *et al.*[11] observed that acrylamide formation increased in a biscuit model system at a moisture content of 2%, levels were moderate at 6% moisture and no acrylamide formation was observed at 10% moisture. However, it is well known that moisture is related to thermal processing conditions and it is difficult to separate the effect of water content from that of temperature or time of heating on acrylamide formation.[101]

Oven type and different baking processes will determine the heat and mass transfer activity, causing the formation of a temperature gradient within the product and, consequently, affecting the rates of acrylamide development in the baked product.[102] In this regard, baking in deck ovens, where heat transfer occurs *via* conduction and radiation, reduces acrylamide levels in bakery products. In contrast, baking in a convection oven, where air circulation is forced and as a consequence the product dries more intensely, promotes higher acrylamide formation.[7] On the other hand, vacuum baking transfers heat under low pressure, which promotes a lower time–temperature profile and results in lower acrylamide content than that attained during conventional oven cooking processes.[100] Lower acrylamide levels have also been observed in cookies,[103] muffins[104] and white bread[101] baked in a steam-assisted oven compared with those baked in convection ovens. The use of steam during baking increases the water content and decreases temperature in the product, which avoids acrylamide formation. An alternative is the application of radiofrequency heating in the last stages of the baking. Radiofrequency allows rapid and uniformly heated foods, reducing the residual moisture of the product and the baking times. The application of radiofrequency to biscuits together with a conventional oven have shown acrylamide concentrations 35% lower than those obtained by using only conventional heating.[105] Regarding breakfast cereals production, extrusion techniques can be crucial for acrylamide formation. Comparing different extrusion cooking methods, the higher thermal input during direct-expansion extrusion gives rise to high acrylamide levels as compared with pellet-to-flaking extrusion.[106] In contrast, very low acrylamide levels, usually below 100 $\mu g\ kg^{-1}$, have also been reported in breakfast cereals made by extrusion puffing.[107]

3.4.3.2 Fermentation

Fermentation can be an effective way of reducing the formation of acrylamide in bread since the addition of yeast to the dough of leavened bread can reduce the content of reducing sugars and asparagine. Decreases of 40–90% of free asparagine have been reported after yeast fermentation,[7,32,108] which would have a greater effect on acrylamide formation during subsequent baking. Fermentation time may also have a strong impact on this strategy. Thereby, prolonging this time up to 1 h has been described as a way to decrease acrylamide levels in industrial bread production. However, longer times (close to 3 h) are not recommended: on the one hand because the yeast's capacity for asparagine metabolization is limited and, on the other hand, due to degradation of the gluten network that triggers the flattening of bread. Fredriksson *et al.*[108] studied the content of asparagine in milling fractions of wheat and rye and the effect of fermentation from 0 to 6 h. Results revealed that crust from breads with wholegrain wheat contained 180 μg acrylamide kg^{-1} when fermented for a short period of time and only 24 μg kg^{-1} when fermented for the longer time. The same treatments in rye bread produced values of 50 μg kg^{-1} and 12 μg kg^{-1}, respectively.

A further approach for acrylamide reduction in bakery products is fermentation by using lactic acid bacteria (LAB), which may consume asparagine or reducing sugars before the food processing step. Acrylamide decreases in crispbread[109] and rye bread have been described when LAB were added during preparation. The effect occurs fundamentally by decreasing the pH but also by reduction of the levels of reducing sugars in the dough. Nevertheless, temperature, pH and time need to be controlled in order to maximize the activity of the LAB microorganisms. For instance, an endogenous proteolytic activity could cause an increase in the asparagine content if fermentation time is prolonged, leading to elevated levels of acrylamide in the baked product.[109]

3.4.3.3 Finished Product: Colour, Texture and Flavour

It is necessary to take into account that acrylamide is generated as a consequence of the Maillard reaction but is also responsible for the production of colour and flavour which are essential for bakery products. For that reason, it is recommended that manufacturers consider baking to a lighter colour endpoint in the final product but achieving the required quality, shelf life, sensory properties and food safety standards. On the other hand, products are often baked at high temperatures and low final moisture content in order to have a 'crisp' texture, which leads to higher levels of acrylamide. Increasing the moisture specification for the final product if feasible for the particular product is recommended.[20]

3.4.4 Final Preparation

Following the recommendations reported in the acrylamide toolbox for fine bakery wares, product pre-mixes prepared for baking at home or at catering establishments should include clear preparation instructions to ensure acrylamide levels are maintained below the benchmark values.

3.5 Additional Considerations and Future Trends

Toxicokinetic studies have described that acrylamide is rapidly and efficiently absorbed in the body due to its low molecular weight and high hydro-solubility.[110] Acrylamide is extensively detoxified by glutathione conjugation and subsequently degraded to the corresponding mercapturic acid that is excreted in urine.[111] A minority fraction (≈1%) is metabolized by cytochrome P450 to glycidamide, the metabolite able to form DNA adducts.[110] Despite this evidence, few studies have progressed our understanding of the effect on the digestive process of acrylamide bioaccesibility and its role in controlling acrylamide toxicity, since 100% bioavailability is assumed.[112] However, the mechanical action during chewing and digestion itself (pH changes and enzymatic activity) could affect its chemical stability/integrity. The high reactivity of this molecule makes possible its participation in additional reactions with nucleophilic groups such as SH– or NH_2– from biomolecules.[113] In short, the presence of acrylamide in food is the result of the balance between the formation of the monomer and its elimination, since it is not a final product.

Hamzalioglu and Gökmen[114] describe that acrylamide content in the bioaccessible fraction of biscuits and some potato-based fried products evolves during the digestive process and is decreased to some extent once the digestion process finishes. Other researchers have documented no differences in acrylamide bioaccesibility from meals such as French fries, gingerbread and its crust compared with an acrylamide solution.[111] Sansano *et al.*[115] have shown that the gastric phase of digestion triggers the release of acrylamide from the food matrix. A combination of several hypotheses can explain this finding:

1. the gastric pH in the stomach promotes *de novo* synthesis from intermediate products of the initial Maillard reaction (Schiff base);[114]
2. the mechanical action and the proteolytic activity by pepsin degrade the food matrix and release the contaminant from the internal core;
3. acid pH can also release acrylamide linked to functionalized macromolecules such as melanoidins.[116]

At the beginning of the intestinal phase, acrylamide seems to decrease, probably due to interaction with small nucleophilic peptides (coming from the action of pepsin) through the Michael addition reaction to form adducts (Figure 3.3) that apparently decrease acrylamide bioaccesibility[115] and consequently its toxic activity. Something similar has also been described for carbonyl compounds such as acrolein, glyoxal, methylglyoxal, HMF and malondialdehyde, whose toxicity against Caco-2 cells was decreased by the formation of adducts with amino acids derived from protein digestion, especially cysteine.[117]

Another interesting aspect is the potential action of the intestinal microbiota on the non-bioaccessible acrylamide reaching the colon. If a portion of the compound was released here it would be available to be absorbed or exert *in situ* local actions. In fact, Bergheim *et al.*[118] established the ability of different colonic segments to express the specific cytochrome P450 (CYP2E1) which metabolizes acrylamide into glycidamide. On the other hand, at this level, toxicity could be diminished if the compound was hydrolyzed *in situ* to non-harmful products. Various microorganisms are able to degrade acrylamide, bacteria being more efficient than yeasts.[119] This is possible thanks to some type of amidase enzymes, which catalyze hydrolysis of the compound to ammonia and acrylic acid. Some microorganisms produce amidases that have demonstrated high substrate specificity, whereas others synthesize amidases able to work on various substrates.[119]

Therefore, the scarcity of information on these mechanisms indicates that the behavior of acrylamide in the gastrointestinal tract, including the digestive process and microbiota action, is largely dependent on the food matrix and the intestinal microbiome and should be studied further. Perhaps human beings have their own internal mitigation strategies in order to decrease *in vivo* acrylamide toxicity.

References

1. T. Y. Curtis, J. Postles and N. G. Halford, *J. Cereal Sci.*, 2014, **59**, 382.
2. R. H. Stadler, I. Blank, N. Varga, F. Robert, J. Hau, P. A. Guy, M. C. Robert and S. Riediker, *Nature*, 2002, **419**, 449.
3. M. Granvogl and P. Schieberle, *J. Agric. Food Chem.*, 2006, **54**, 5933.
4. A. Claus, G. M. Weisz, A. Schieber and R. Carle, *Mol. Nutr. Food Res.*, 2006, **50**, 87.
5. V. A. Yaylayan, A. Wnorowski and C. Perez Locas, *J. Agric. Food Chem.*, 2003, **51**, 1753.
6. D. V. Zyzak, R. A. Sanders, M. Stojanovic, D. H. Tallmadge, B. L. Eberhart, D. K. Ewald, D. C. Gruber, T. R. Morsch, M. A. Strothers, G. P. Rizzi and M. D. Villagran, *J. Agric. Food Chem.*, 2003, **51**, 4782.
7. A. Claus, R. Carle and A. Schieber, *J. Cereal Sci.*, 2008, **47**, 118.
8. A. Mustafa, P. Aman, R. Andersson and A. Kamal-Eldin, *Food Chem.*, 2007, **105**, 317.
9. S. Žilić, D. Dodig, Z. Basić, J. Vančetović, P. Titan, N. Đurić and N. Tolimir, *Food Addit. Contam., Part A*, 2017, **34**, 705.
10. T. Y. Curtis, S. J. Powers, R. Wang and N. G. Halford, *Food Chem.*, 2018, **239**, 304.
11. D. Taeymans, J. Wood, P. Ashby, I. Blank, A. Studer, R. H. Stadler, P. Gonde, P. Van Eijck, S. Lalljie, H. Lingnert, M. Lindblom, R. Matissek, D. Muller, D. Tallmadge, J. O'Brien, S. Thompson, D. Silvani and T. Whitmore, *Crit. Rev. Food Sci. Nutr.*, 2004, **44**, 323.
12. T. Y. Curtis, N. Muttucumaru, P. R. Shewry, M. A. J. Parry, S. J. Powers, J. S. Elmore, D. S. Mottram, S. Hook and N. G. Halford, *J. Agric. Food Chem.*, 2009, **57**, 1013.
13. J. S. Elmore, G. Koutsidis, A. T. Dodson, D. S. Mottram and B. L. Wedzicha, *J. Agric. Food Chem.*, 2005, **53**, 1286.
14. M. Przygodzka, M. K. Piskula, K. Kukurová, Z. Ciesarová, A. Bednarikova and H. Zieliński, *J. Cereal Sci.*, 2015, **65**, 96.

15. P. J. Lea, L. Sodek, M. A. J. Parry, R. Shewry and N. G. Halford, *Ann. Appl. Biol.*, 2007, **150**, 1.
16. K. N. Englyst, S. Liu and H. N. Englyst, *Eur. J. Clin. Nutr.*, 2007, **61**, S19.
17. J. I. Mann and J. H. Cummings, *Nutr., Metab. Cardiovasc. Dis.*, 2009, **19**, 226.
18. P. Vitaglione, A. Napolitano and V. Fogliano, *Trends Food Sci. Technol.*, 2008, **19**, 451.
19. L. Nyström, A. M. Lampi, A. A. M. Andersson, A. Kamal-Eldin, K. Gebruers, C. M. Courtin, J. A. Delcour, L. Li, J. L. Ward, A. Fras, D. Boros, M. Rakszegi, Z. Bedo, P. R. Shewry and V. Piironen, *J. Agric. Food Chem.*, 2008, **56**, 9758.
20. FDE (FoodDrinkEurope), *The Acrylamide Toolbox*, Food Drink Europe, 2014, p. 57.
21. EFSA (European Food Safety Authority), *EFSA J.*, 2015, **13**, 4104.
22. EC (European Commission), *Off. J. Eur. Union*, 2017, **L304**, 24.
23. FAO (Food and Agriculture Organization), *FAO Cereal Supply and Demand Brief report*, 2018, Access 16/12/2018, http://www.fao.org/worldfoodsituation/csdb/en/.
24. C. Molina-Rosell, in *Flour and Breads and Their Fortification in Health and Disease Prevention*, ed. V. R. Preedy, R. R. Watson and V. B. Patel, Academic Press, London, UK, 2011.
25. EFSA (European Food Safety Authority), *EFSA J.*, 2011, **9**, 2133.
26. M. Palermo, V. Gokmen, B. De Meulenaer, Z. Ciesarova, Y. Zhang, F. Pedreschi and V. Fogliano, *Food Funct.*, 2016, **7**, 2516.
27. N. Muttucumaru, N. G. Halford, S. J. Elmore, A. T. Dodson, M. Parry, P. P. Shewry and D. S. Mottram, *J. Agric. Food Chem.*, 2006, **54**, 8951.
28. F. Stockmann, E. A. Weber, B. Mast, P. Schreiter, N. Merkt, W. Claupein and S. Graeff-Hönninger, *Agronomy*, 2018, **8**, 294.
29. Zh. S. Kudiyarova, N. Muttucumaru, J. West and N. Halford, *Asp. Appl. Biol.*, 2013, **116**, 119.
30. E. Capuano, G. Garofalo, A. Napolitano, H. Zielinski and V. Fogliano, *Cereal Chem.*, 2010, **87**, 131.
31. M. Springer, T. Fischer, A. Lehrack and W. Freund, *Getreide, Mehl Brot*, 2003, **57**, 274.
32. S. Wang, J. Yu, Q. Xin, S. Wang and L. Copeland, *Food Control*, 2017, **73**, 230.
33. EU (European Commission), https://ec.europa.eu/jrc/en/health-knowledge-gateway/promotion-prevention/nutrition/fibre#_Tocch3, accessed December 2018.
34. K. A. Harris and P. M. Kris-Etherton, *Curr. Atheroscler. Rep.*, 2010, **12**, 368.
35. W. Guo and T. Beta, *Food Res. Int.*, 2013, **51**, 518.
36. P. Vitaglione, A. Napolitano and V. Fogliano, *Trends Food Sci. Technol.*, 2008, **18**, 451.
37. W. C. Cheng, D. C. Sun, S. S. Chou and A. I. Yeh, *J. Food Prot.*, 2012, **75**, 2158.

38. K. H. Liukkonen, R. L. Heinio, M. Salmenkallio-Marttila, K. Autio, K. Katina and K. Poutanen, Rye, in *Bakery Products: Science and Technology*, ed. W. Zhou and Y. H. Hui, Blackwell Publishing, Oxford, UK, 2006, pp. 109–122.
39. W. Bushuk, *Rye: Production, Chemistry, and Technology*, American Association of Cereal Chemists, St. Paul, MN, 2001.
40. E. Capuano, A. Ferrigno, I. Acampa, A. Serpen, O. C. Acar, V. Gokmen and V. Fogliano, *Food Res. Int.*, 2009, **42**, 1295.
41. C. Delgado-Andrade, M. Navarro, H. López and M. C. López, *Food Addit. Contam.*, 2003, **20**, 923.
42. C. G. Hamlet, P. A. Sadd, L. Liang, S. M. Jayaratne and M. Skingle, Exploiting processing conditions to reduce acrylamide in cereal-based foods, *RHM Technology*, Report No. C021, 2007..
43. M. Graf, T. M. Amrein, S. Graf, R. Szalay, F. Escher and R. Amadò, *LWT--Food Sci. Technol.*, 2006, **39**, 724.
44. M. Vass, T. M. Amrein, B. Schönbächler, F. Escher and R. Amadò, *Czech J. Food Sci.*, 2004, **22**, 19.
45. V. Gökmen, Ö. Ç. Açar, H. Köksel and J. Acar, *Food Chem.*, 2007, **104**, 1136.
46. H. T. Nguyen, H. J. I. van der Fels-Klerx and M. A. J. S. van Boekel, *Food Chem.*, 2017, **1**, 230.
47. N. Surdyk, J. Rosén, R. Andersson and P. Åman, *J. Agric. Food Chem.*, 2004, **52**, 2047.
48. M. Beck, M. Jekle and T. Becker, *J. Sci. Food Agric.*, 2012, **92**, 585.
49. E. Kolek, P. Simko and P. Simon, *Food Nutr. Res.*, 2006, **45**, 17.
50. H. J. Van Der Fels-Klerx, E. Capuano, H. T. Nguyen, B. A. Mogol, T. Kocadagli, N. Goncuoglu Tas, A. Hamzalioglu, M. A. J. S. van Boekel and V. Gokmen, *Food Res. Int.*, 2014, **57**, 210.
51. WHO (Word Health Organization), Guideline: sodium intake for adults and children, 2012, WHO Document Production Services, Geneva, Switzerland.
52. J. Powles, S. Fahimi, R. Micha, S. Khatibzadeh, P. Shi, M. Ezzati, R. E. Engell, S. S. Lim, G. Danaei and D. Mozaffarian, *BMJ Open*, 2013, **3**, e003733.
53. S. A. O'Halloran, C. A. Grimes, K. E. Lacy, C. A. Nowson and K. J. Campbell, *BMJ Open*, 2016, **6**, e008698.
54. E. Capuano, T. Oliviero, Ö. Ç. Açar, V. Gökmen and V. Fogliano, *Food Res. Int.*, 2010, **43**, 1021.
55. G. Arribas-Lorenzo, V. Fogliano and F. J. Morales, *Eur. Food Res. Technol.*, 2009, **229**, 63.
56. R. Zamora, E. Gallardo and F. J. Hidalgo, *J. Agric. Food Chem.*, 2007, **55**, 1308.
57. H. M. Lai and T. C. Lin, in *Bakery Products. Science and Technology*, ed. H. Y. Hui, Blackwell Publishing Professional, Iowa, USA, 2006, pp. 3–65.
58. K. Kukurová, Z. Ciesarova, B. A. Mogol and V. Gökmen, *Eur. Food Res. Technol.*, 2013, **237**, 1.

59. O. Marconi, E. Bravi, G. Perreti, R. Martini, L. Montanari and P. Fantozzi, *Anal. Methods*, 2010, **2**, 1686.
60. T. M. Amrein, L. Andres, G. G. G. Manzardo and R. Amado, *J. Agric. Food Chem.*, 2006, **54**, 10253.
61. R. A. Levine and R. E. Smith, *J. Agric. Food Chem.*, 2005, **53**, 4410.
62. T. M. Amrein, B. Schönbächler, F. Escher and R. Amado, *J. Agric. Food Chem.*, 2004, **52**, 4282.
63. D. Manley, in *Manley's Technology of Biscuits, Crackers and Cookies*, ed. D. Manley, Woodhead Publishing, Cambridge, UK, 2011, pp. 223–234.
64. E. Capuano, A. Ferrigno, I. Acampa, L. Ait-Ameur and V. Fogliano, *Eur. Food Res. Technol.*, 2008, **228**, 311.
65. E. B. Brathen, A. Kita, S. H. Knutsen and T. Wicklund, *J. Agric. Food Chem.*, 2005, **53**, 3259.
66. M. Y. Low, J. K. Parker and D. S. Mottram, *J. Agric. Food Chem.*, 2007, **55**, 4087.
67. R. Flückiger and E. Salih, Method to limit acrylamide in heated foods, *W.O. Pat.*, WO2006017526, 2006.
68. P. A. Sadd, C. G. Hamlet and G. L. Liliang, *J. Agric. Food Chem.*, 2008, **56**, 6154.
69. V. A. Elder, J. G. Fulcher, H. Leung and M. G. Topor, Method for reducing acrylamide in thermally processed foods, *U.S. Pat.*, US20040058045, 2004.
70. R. C. Lindsay and S. Jang, *Adv. Exp. Med. Biol.*, 2005, **561**, 393.
71. M. Mesías, F. Holgado, G. Márquez-Ruiz and F. J. Morales, *LWT--Food Sci. Technol.*, 2015, **62**, 633.
72. R. A. Levine and S. M. Ryan, *J. Agric. Food Chem.*, 2009, **57**, 6823.
73. H. V. Hendriksen, B. A. Kornbrust, P. R. Østergaard and M. A. Stringer, *J. Agric. Food Chem.*, 2009, **57**, 4168.
74. N. B. Tuncel, N. Yılmaz and E. Sener, *Int. J. Food Sci. Technol.*, 2010, **45**, 2470.
75. M. Anese, B. Quarta, L. Peloux and S. Calligaris, *Food Res. Int.*, 2011, **44**, 2837.
76. Z. Ciesarová, K. Kukurová, L. Mikušová, E. Basil, P. Polakovičová, L. Duchoňová, M. Vlček and E. Šturdík, *Qual. Assur. Saf. Crops Foods*, 2014, **6**, 327.
77. N. S. M. Kumar, C. A. Shimray, D. Indrani and H. K. Manonmani, *Food Bioprocess Technol.*, 2014, **7**, 741.
78. L. Huang, Y. Liu, Y. Sun, Q. Yan and Z. Jiang, *Appl. Environ. Microbiol.*, 2014, **80**, 1561.
79. T. Thompson, *J. Am. Diet. Assoc.*, 2001, **101**, 586.
80. R. Kamboj and V. Nanda, *Legume Res.*, 2018, **41**, 325.
81. J. H. Y. Galani, N. J. Patel and J. G. Talati, *Food Chem. Toxicol.*, 2017, **108**, 244.
82. S. Milkovska-Stamenova, L. Krieg and R. Hoffmann, *Mol. Nutr. Food Res.*, 2018, 1800725.
83. D. Rachwal-Rosiak, E. Nebesny and G. Budryn, *Crit. Rev. Food Sci. Nutr.*, 2015, **55**, 1135.

84. P. Rozan, Y. H. Kuo and F. Lambein, *Phytochemistry*, 2001, **58**, 281.
85. L. Sodek, P. J. Lea and B. J. Miflin, *Plant Physiol.*, 1980, **65**, 22.
86. L. Sodek and P. J. Lea, *Phytochemistry*, 1993, **34**, 51.
87. Y. Cai, K. C. Sam Chang, S. Y. Ou and Z. Liu, *Mod. Food Sci. Technol.*, 2014, **30**, 37.
88. F. Pedreschi, I. Saavedra, A. Bunger, R. N. Zuñiga, R. Pedreschi, R. Chirinos, D. Campos and M. S. Mariotti-Celis, *LWT--Food Sci. Technol.*, 2018, **95**, 116.
89. M. Mesías, C. Delgado-Andrade and F. J. Morales, *J. Sci. Food Agric.*, 2018, **99**, 2048.
90. E. Reyes-Caudillo, A. Tecante and M. A. Valdivia-López, *Food Chem.*, 2008, **107**, 656.
91. L. Alvarez-Jubete, E. K. Arendt and E. Gallaghera, *Trends Food Sci. Technol.*, 2010, **21**, 106.
92. M. Mesías, F. Holgado, G. Márquez-Ruiz and F. J. Morales, *LWT--Food Sci. Technol.*, 2016, **73**, 528.
93. M. Mesías and F. J. Morales, *Foods*, 2017, **6**, 14.
94. R. Salazar, G. Arámbula-Villa, P. A. Vázquez-Landaverde, F. J. Hidalgo and R. Zamora, *Food Chem.*, 2012, **135**, 2293–2298.
95. L. Alvarez-Jubete, E. K. Arendt and E. Gallagher, *Int. J. Food Sci. Nutr.*, 2009, **60**, 240.
96. P. R. Shewry, *J. Cereal Sci.*, 2018, **79**, 469.
97. Y. Zhang and Y. Zhang, *Asia Pac. J. Clin. Nutr.*, 2007, **16**, 131.
98. R. V. Hedegaard, K. Granby, H. Frandsen, J. Thygesen and L. H. Skibsted, *Eur. Food Res. Technol.*, 2008, **227**, 519.
99. O. C. Açar and V. Gökmen, *Mol. Nutr. Food Res.*, 2009, **53**, 1521.
100. B. A. Mogol and V. Gökmen, *Innovative Food Sci. Emerging Technol.*, 2014, **26**, 265.
101. L. Ahrné, C. G. Andersson, P. Floberg, J. Rosén and H. Lingnert, *LWT--Food Sci. Technol.*, 2007, **40**, 1708.
102. O. C. Açar and V. Gökmen, *J. Food Eng.*, 2010, **100**, 642.
103. H. Isleroglu, T. Kemerli, M. Sakin-Yilmazer, G. Guven, O. Ozdestan, A. Uren and F. Kaymak-Ertekin, *J. Food Sci.*, 2012, **77**, E257.
104. M. Sakin-Yilmazera, T. Kemerli, H. Isleroglu, O. Ozdestan, G. Guven, A. Uren and F. Kaymak-Ertekin, *J. Food Eng.*, 2013, **119**, 483.
105. M. Anese, S. Sovrano and R. Bortolomeazzi, *Eur. Food Res. Technol.*, 2008, **226**, 1197.
106. J. A. Rufián-Henares, C. Delgado-Andrade and F. J. Morales, *Mol. Nutr. Food Res.*, 2006, **50**, 756.
107. R. H. Stadler, in *Acrylamide and Other Hazardous Compounds in Heat-treated Foods*, ed. K. Skog and J. Alexander, Woodhead Publishing, Cambridge, UK, 2006, pp. 23–40.
108. H. Fredriksson, J. Tallving, J. Rosen and P. Aman, *Cereal Chem.*, 2004, **81**, 650.
109. P. Baardseth, H. Blom, G. Enersen, G. Skrede, E. Slinde, T. Sundt and T. Thomassen, Reduction of acrylamide formation in cereal-based food processing, *W.O. Pat.*, WO2004028276, 2004.

110. M. Friedman, *Food Funct.*, 2015, **6**, 1752.
111. F. I. Berger, J. Feld, D. Bertow, G. Eisenbrand, G. Fricker, N. Gerhardt, K. H. Merz, E. Richling and M. Baum, *Mol. Nutr. Food Res.*, 2011, **55**, 387.
112. S. Eriksson, *LWT--Food Sci. Technol.*, 2005, **39**, 393.
113. R. H. Stadler, L. Verzegnassi, N. Varga, M. Grigorov, A. Studer, S. Riediker and B. Schilter, *Chem. Res. Toxicol.*, 2003, **16**, 1242.
114. A. Hamzalioglu and V. Gokmen, *Food Funct.*, 2015, **6**, 108.
115. M. Sansano, A. Heredia, I. Peinado and A. Andrés, *Food Chem.*, 2017, **237**, 58.
116. S. Pastoriza, J. A. Rufián-Henares and F. J. Morales, *LWT--Food Sci. Technol.*, 2012, **45**, 198.
117. K. Jiang, C. Huang, R. Jiao, W. Bai, J. Zheng and S. Ou, *J. Hazard. Mater.*, 2019, **363**, 26.
118. I. Bergheim, C. Bode and A. Parlesak, *BMC Clin. Pharmacol.*, 2005, **5**, 4.
119. A. Duda-Chodak, L. Wajda, P. Sroka and P. Satora, *Food Funct.*, 2016, **7**, 1282.

CHAPTER 4

2- and 3-Monochloropropanediol (MCPD) Esters and Glycidyl Esters: Methods of Analysis, Occurrence, and Mitigation in Refined Oils, Infant Formula, and Other Processed Foods

JESSICA K. BEEKMAN*[a], MICHAEL GRANVOGL[b,c] AND SHAUN MACMAHON[a]

[a]U.S. Food and Drug Administration, Center for Food Safety and Applied Nutrition (CFSAN), 5001 Campus Drive, College Park, MD 20740, USA; [b]Technical University of Munich, Chair for Food Chemistry, Lise-Meitner-Straße 34, 85354 Freising, Germany; [c]Institute of Food Chemistry, University of Hohenheim, Garbenstraße 28, 70599 Stuttgart, Germany
*E-mail: jessica.beekman@fda.hhs.gov

4.1 Introduction

3-Monochloropropanediol (3-MCPD) and 2-monochloropropanediol (2-MCPD) belong to a class of environmental contaminants known as chloropropanols. These compounds possess the basic glycerol structure, but with a chlorine atom replacing a hydroxyl group in either the sn-1 (3-MCPD) or sn-2 (2-MCPD) position (Figure 4.1).

Food Chemistry, Function and Analysis No.19
Mitigating Contamination from Food Processing
Edited by Catherine S. Birch and Graham A. Bonwick
© The Royal Society of Chemistry 2020
Published by the Royal Society of Chemistry, www.rsc.org

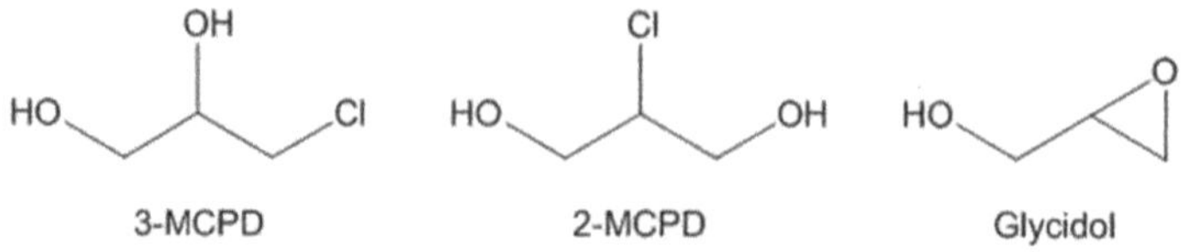

Figure 4.1 Structures of 3-MCPD, 2-MCPD, and glycidol.

These contaminants, which are typically formed as undesirable by-products during the production of various foods and food ingredients, were first discovered in the late 1970s in acid-hydrolyzed vegetable protein (acid-HVP).[1–3] Acid-HVP, which is used to enhance the "meaty" flavor in a variety of foodstuffs such as soups, sauces, and chili, is prepared *via* the hydrolysis of protein (typically from sources of vegetable origin) with hydrochloric acid (HCl) at elevated temperatures. However, during this preparation HCl can react with the glycerol backbone of the residual lipid content from the protein source, resulting in the formation of 3-MCPD, and to a lesser extent 2-MCPD.[4,5] In addition to acid-HVP, these contaminants were also discovered at elevated levels in the 1990s and early 2000s in some Asian sauces, such as oyster sauce, hoisin sauce, and soy sauce.[6–8] When prepared traditionally *via* soybean fermentation, these sauces can take up to 6 months to complete. However, this production time can be reduced to a matter of days when fermentation is replaced with acid hydrolysis, involving prolonged heating at elevated temperatures to break down the soy protein. Unfortunately, the use of chlorinated acid during the hydrolysis process results in the production of MCPD *via* the reaction of chloride with residual lipid components.[4] Studies have also revealed the presence of MCPD in a variety of other foodstuffs such as bread, coffee, cheese, meat, salted fish, and smoked food products.[9–11] Although the mechanisms of MCPD formation are not well known in many of these food products, these contaminants are believed to arise from the reaction of chlorine-containing species with precursors naturally present in these foods.[9,12]

In addition to free 2- and 3-MCPD, fatty acid esters of these contaminants were also reported in acid-HVP,[2,3] although it wasn't until the mid-2000s that researchers began extensively investigating the occurrence of these compounds in other foodstuffs.[13–18] These contaminants possess the basic 2- and 3-MCPD backbone (with a chlorine in the sn-2 or sn-1 position), but with one (2- and 3-MCPD monoesters) or two (2- and 3-MPCD diesters) fatty acid chains bonded to oxygens of the hydroxyl groups to form esters. Figure 4.2 illustrates a few examples of these fatty acid ester contaminants, with the actual number of possible fatty acid ester combinations (with fatty acid carbon chain lengths ranging from 4 to 24) totaling well over 100.

With the discovery of 2- and 3-MCPD esters in acid-HVP and other foodstuffs, it was hypothesized that these contaminants would also be ubiquitous

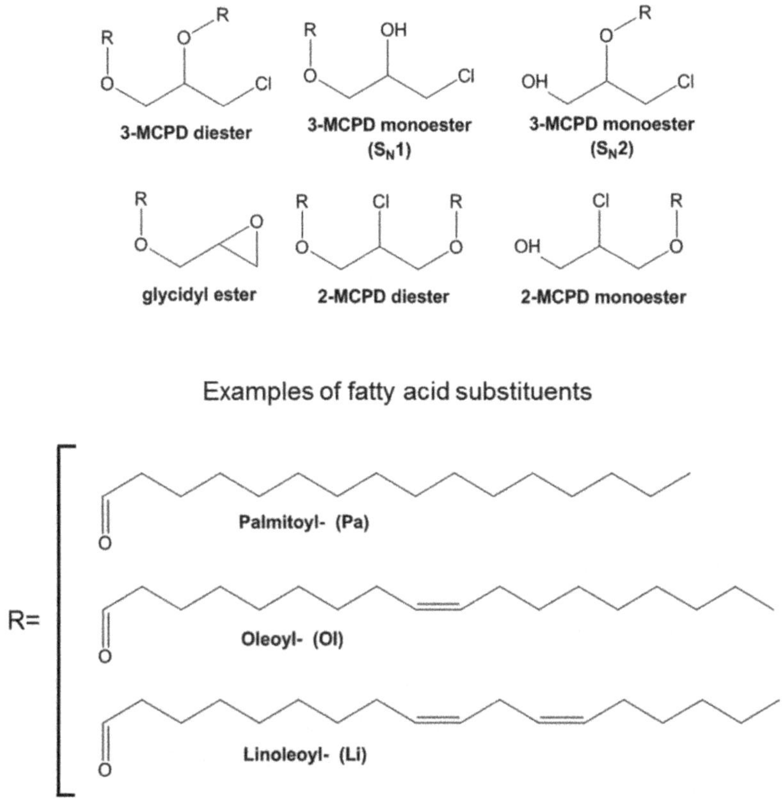

Figure 4.2 Structures of fatty acid esters of 3-MCPD, 2-MCPD, and glycidol and examples of possible fatty acid substituents.

in refined edible oils due to processing conditions and the characteristic components of vegetable fats and oils.[13] In order to improve consumer acceptance, the vast majority of edible fats and oils undergo processing to remove components that can negatively impact taste, smell, appearance, nutritional content, and shelf stability, and for the elimination of toxicologically relevant compounds such as pesticides and heavy metals. The deodorization step of the refining process is particularly effective at reducing or eliminating these undesirable components, but studies have confirmed that the elevated temperatures required for this stage of refining contribute to the formation of fatty acid esters of 2- and 3-MCPD. Since this discovery, a number of methods have been developed for the analysis of these contaminants in fats and oils, and occurrence data for these compounds in edible oils from all over the world have been published.[13–15,19–25]

In addition to fatty acid esters of 2- and 3-MCPD, studies in the late 2000s revealed the presence of previously undiscovered fatty acid ester contaminants of glycidol in refined oils.[26–28] These esters possess the basic glycidol backbone (Figure 4.1, containing the epoxide and alcohol functional groups), but, similar to the MCPD monoesters, contain one fatty acid chain bonded to the oxygen of the hydroxyl group (Figure 4.2). Unlike free 2- and 3-MCPD, free glycidol is not found in foods and, because of its relative instability, it is rarely encountered in pure form. However, in esterified form, numerous studies have shown glycidyl ester contaminants can be found at elevated levels in nearly all refined edible oils.[20,24,29–32]

The presence of 2- and 3-MCPD esters and glycidyl esters has also been reported in a number of food products. Many processed foods, such as French fries, noodles, chips, cookies, cakes and candies, contain refined edible oils as ingredients, which consequently results in the introduction of these processing contaminants into the final product. Recently, specific emphasis has been placed on the evaluation of MPCD and glycidyl ester levels in infant formulas. Because refined vegetable oils are the primary fat source in commercial infant formulas, and because formula could be an infant's sole source of nutrition, there has been some level of concern regarding infants' exposure to these contaminants. These concerns have prompted the recent and on-going development of methods for the extraction and quantitative determination of 2- and 3-MCPD and glycidyl ester levels in infant formula, as well as other processed foods. Although a few single-laboratory validated methods have been published,[33–35] at present there is only one collaboratively studied official method for the analysis of MCPD and glycidyl esters in dressings and spreads containing refined oils.[36]

In February 2018, the European Commission published an updated regulation (EC 2018/290) implementing maximum allowable levels of glycidyl esters (reported as bound glycidol) in refined edible oils and infant formulas.[37] Bound glycidol concentrations in vegetable oils and fats intended for sale to the consumer or to be used in food production cannot exceed 1 mg kg^{-1}, and bound glycidol levels in oils/fats intended for use in the production of infant formulas cannot exceed 0.5 mg kg^{-1}. This regulation has also specified that powdered and liquid infant formula products cannot contain bound glycidol levels exceeding 75 μg kg^{-1} and 10 μg kg^{-1}, respectively, and that these maximum allowable levels will be reduced to 50 μg kg^{-1} and 6 μg kg^{-1} respectively, beginning in July of 2019. Although the regulation specifies limits for free 3-MCPD in acid-HVP and soy sauces (20 μg kg^{-1}), there is currently no guidance on maximum allowable levels of bound 3-MCPD in refined edible oils and infant formula. The European Food Safety Authority (EFSA) has recently revised its risk assessment of free 3-MCPD and its esters,[38] and it is likely the results of this report will lead to an update of the regulation to include maximum limits for bound 3-MCPD in oils/fats and infant formulas in the near future.

4.2 Toxicology

Numerous studies have reported that free 2- and 3-MCPD and glycidol may pose a number of health concerns.[39–42] In addition, research has shown that fatty acid esters of these contaminants undergo de-esterification in the gastrointestinal tract, releasing free 2-/3-MCPD and glycidol with a high degree of efficiency.[43,44] Therefore, toxicity of the esters is typically evaluated based on conversion of fatty acid ester concentrations into free MCPD and glycidol equivalents.

Free 3-MCPD has been classified as a group 2B carcinogen (possibly carcinogenic to humans) by the International Agency for Research on Cancer (IARC).[45] In short-term studies, repeated doses as low as 5.2 mg 3-MCPD/kg body weight daily in rats resulted in renal toxicity, with neurotoxic effects (such as limb paralysis) observed at doses over 50 mg 3-MCPD/kg body weight daily.[46] In addition to renal toxicity, long-term dosing studies revealed that daily doses as low as 2 mg 3-MCPD/kg body weight in rats also caused testicular toxicity, with slightly higher doses severely impairing male fertility.[47] In 2017, the Joint Expert Committee on Food Additives (JECFA) recommended a tolerable daily intake (TDI) of 4 μg 3-MCPD/kg body weight daily,[48] and in 2018 the EFSA published an updated risk assessment of 3-MCPD and its esters establishing a TDI of 2 μg 3-MCPD/kg body weight daily.[38]

Glycidol has been classified as a Group 2A carcinogen (probably carcinogenic to humans) by the IARC, with a number of studies providing evidence of its ability to act as a direct-alkylating agent (genotoxicity).[49] In short-term studies, nephrotoxicity resulted in mice and rats at 150–400 mg glycidol/kg body weight doses for 13 weeks,[50] and neurotoxic effects were observed in rats following doses of 200 mg glycidol/kg body weight when administered daily for 28 days.[51] In long-term studies (approximately 2 years), the development of tumors in multiple organs, likely through a genotoxic mode of action, was observed in both mice and rats at doses of between 25 and 75 mg glycidol/kg body weight administered 5 days per week. Reproductive toxicity was also observed in male mice and rats at doses as low as 75 and 25 mg glycidol/kg body weight daily, respectively, when administered for 13 weeks.[50] Although a TDI has not been recommended for glycidol, the Food Safety Commission of Japan has suggested glycidol intake should be based on the ALARA principle (As Low As Reasonably Achievable) given its genotoxic carcinogenicity.[52]

There have been limited toxicological studies involving free 2-MCPD, which has made thorough evaluation of its potential carcinogenicity difficult. In an unpublished short-term study, severe myopathy of striated muscle tissue (particularly cardiac muscle tissue) and nephrotoxicity were observed in rats at doses of 16 or 30 mg 2-MCPD/kg body weight daily.[53] At day 8 of the study, development of lesions in striated muscle (including the heart) could be observed, which resulted in heart failure and death of some animals. It was also noted that renal toxicity was not observed in rats that were administered doses of 2-MCPD (~5 mg kg^{-1} body weight daily) equivalent to 3-MCPD doses that produced renal failure. To date, long-term toxicity studies for 2-MCPD have not been performed and its potential genotoxicity has not been evaluated. Given the limited data available, there is currently no health-based guidance for maximum daily intake of 2-MCPD.

Recently, two studies have been published investigating the possible toxicity of several fatty acid esters of 3-MCPD.[54,55] In short-term studies with mice and rats, Liu *et al.* (2012),[54] observed severe neurotoxic effects such as lateral walking and limb paralysis after administering various doses of 1-palmitoyl-3-chloropropanediol (monoester) and 1,2-bis-palmitoyl-3-chloropropanediol (diester). In addition, toxicity of the thymus and lungs was reported following administration of various doses of 1-stearoyl-3-chloropropanediol, 1-oleoyl-3-chloropropanediol, 1-linoleoyl-3-chloropropanediol, 1-palmitoyl-2-linoleoyl-3-chloropropanediol, and 1-linoleoyl-2-palmitoyl-3-chloropropanediol. Although the authors acknowledged the observed toxic effects may arise from free 3-MCPD following hydrolysis of the esters after ingestion, these studies suggest that fatty acid chain length, saturation, and number of fatty acid substitutions may influence the degree of toxicity. Thus, further investigation of the mechanisms of ester toxicity is necessary.

4.3 Mitigation

Implementation of European Union (EU) regulations for maximum concentrations of bound glycidol in edible oils and infant formulas (and the likely enactment of maximum bound 3-MCPD levels in the EU in the near future) has emphasized the need for strategies to mitigate these ester contaminants. Over the last several years, numerous researchers have investigated various approaches to minimize concentrations of these contaminants in oils and foods, and these methods can be grouped into three potential areas where MCPD and glycidyl esters reduction might be achieved: harvesting, refining, and post-refining. The first two areas involve the reduction of contaminants by limiting their formation in the crude oil, resulting in a "mitigated" refined oil, and the third area refers to the removal of contaminants from the finished product and/or limiting further formation of ester contaminants during the production of foodstuffs containing refined oils. It is important to point out that of the mitigation studies available, the vast majority have been conducted exclusively on palm oil; however, in some cases the conclusions may be applicable across all oils. While the oil industry currently utilizes a number of mitigation approaches for minimizing contaminant levels, many additional strategies, which have currently only been investigated on laboratory or pilot scales, represent immense potential for future industrial use.

A number of researchers have suggested that cultivating and harvesting practices of the oil plants may impact MCPD and glycidyl ester levels in the final oil product. Proposed mitigation approaches include breeding oil plants that contain lower levels of lipase enzymes, which would result in reduced cleavage of triglyceride precursors and, thereby, decreased formation of glycidyl esters during oil refining.[56] Minimizing the introduction of chlorine-containing compounds during cultivation (such as through fertilizers, pesticides, water, *etc.*) could also reduce MCPD ester concentrations in the final oil product since the presence of chlorine species is required for formation during processing.[57,58] Modifications to harvesting and storage practices, such as avoiding bruised or over-ripened fruit (for fruit pulp oils) and minimizing the time between harvesting and milling (which decreases enzymatic triglyceride cleavage), have also been suggested for reducing contaminant levels.[59] While these approaches have shown to be promising in smaller-scale experiments, they may be difficult to implement in practice. Altering cultivation and harvesting procedures could be costly (breeding in particular would be a lengthy process) and minimizing the storage duration of harvested palm fruits could be challenging if plantation fields are considerable distances from milling/refining facilities.

Other potential opportunities for MCPD and glycidyl esters reduction in refined oil products arises from the modification of certain steps during the refining process. Depending on a number of factors, crude oils are either chemically or physically refined, and both processes present opportunities for contaminant reduction. The first step in the refining process involves degumming of the crude oil, which refers to the removal of phospholipids and other phosphorous-containing compounds (or "gums"), by employing various acid/water "washing" procedures depending on the type of oil and gum concentration. A number of studies have shown that modification of the degumming process by either reducing the amount of acid introduced for gum removal or eliminating the use of acids altogether and washing with water at elevated temperatures (approximately 70 °C) can significantly reduce ester levels in the final refined product through removal of chloride ion precursors in the water washes.[60–64] During chemical refining, the degumming process is followed by neutralization of free fatty acids, but this step is not necessarily required for physical refining. Several researchers have shown that the introduction of neutralization agents such as calcium oxide or sodium carbonate can minimize ester formation substantially by removal of precursors.[63,65] However, while introduction of a neutralization step during physical refining could be an effective mitigation strategy, it may also result in corresponding oil losses and large quantities of waste. The bleaching step of the refining process takes place during both chemical and physical refining and involves the removal of color pigments and other trace impurities with porous bleaching clays. A number of studies have demonstrated that the use of neutral bleaching clays, as opposed to acid activated bleaching earths (which are chemically altered to increase bleaching efficiency), can significantly reduce ester formation by maintaining more neutral pH levels during the refining process.[63] At present, there are several companies that produce specialty bleaching clays specifically for minimizing MCPD and glycidyl esters formation during oil refining. Finally, numerous studies have shown that modifications in the deodorization step of the refining process, which involves the final removal of unwanted impurities in the crude oil, have exhibited enormous potential for reducing levels of MCPD and glycidyl esters in refined oils. Decreasing deodorization temperatures to less than 230 °C has been shown to be particularly effective for minimizing the formation of glycidyl esters;[61,66,67] however, this strategy has little effect on MCPD ester levels since their formation begins at approximately 175 °C. Some researchers have demonstrated that a two-stage deodorization process, which involves lower deodorization temperatures (~200 °C) for a longer time period (hours) followed by a boost in deodorization temperature for a shorter time period (minutes), effectively reduces both MCPD and glycidyl ester concentrations.[62] While each of these mitigation strategies has demonstrated effective contaminant reduction, various combinations of these methods may be necessary to achieve the best overall mitigation of ester contaminants in the finished product.

Researchers have also demonstrated the effectiveness of several post-refining mitigation strategies for the removal of MCPD and glycidyl ester contaminants from the final refined product. A number of materials have been investigated for their ability to adsorb the ester contaminants in the refined oil to the surface of the adsorbent substance. Materials such as calcinated zeolite and magnesium silicate have shown promising results in laboratory studies for the removal of glycidyl esters from refined oils.[68] In a recent study, researchers demonstrated that glycidyl ester levels could be reduced by 95% through the use of acid-washed oil palm wood-based activated carbon (OPAC) by adsorbing the contaminants to the material and by degradation of the glycidyl esters at the acid-activated sites.[69] This study did not investigate the impact of OPAC materials on MCPD ester concentrations. Acid-activated bleaching clays have also been shown to effectively reduce glycidyl ester concentrations by 99%, with significant reduction attributed to the acidic conditions that facilitate epoxide ring opening of the glycidyl ester backbone resulting in the transformation of glycidyl esters into acylglycerols and glycerol.[70] While these adsorbent materials have shown immense promise in laboratory experiments, there are still questions about their efficacy in practice due to costs, oil losses, waste products, and the potential impact on oil quality. The use of enzymes to break down the ester contaminants in refined oil products has also been explored. Two studies have demonstrated that conversion of MCPD esters into free MCPD, followed by MCPD conversion into glycidol, and finally degradation of glycidol to form acylglycerols and glycerol could be achieved through enzymatic reactions with the targeted contaminants; however, these studies indicated that further experiments would be necessary to confirm the safety of the degradation products in the final oil product.[70,71] Finally, for some processed food products that contain refined oils the potential exists for further MCPD and glycidyl esters formation during food production. Some studies have suggested that limiting the introduction of chloride-containing salts into the food product and keeping production temperatures low (if possible) may minimize further contaminant formation.[59,72]

Mitigation of MCPD and glycidyl ester contaminants in refined oils remains a challenging and ongoing area of research. Effective mitigation techniques can vary tremendously depending on factors such as the type of oil to be refined and the capabilities of the refinery. Two mitigation "toolboxes", which detail a number of mitigation strategies and their applicability in certain refining situations, have been published to aid industry in the reduction of ester contaminants in their products.[73,74] However, it is evident that there is no "one size fits all approach" for the mitigation of these contaminants in refined oils. It is likely that future studies will encompass investigations into effective combinations of these mitigation strategies as well implementation of such techniques on an industrial scale.

4.4 Methods of Analysis – Vegetable Oils

The discovery of fatty acid esters of MCPD and glycidol in refined vegetable oils has led to the development of numerous methods for the quantitative analysis of these contaminants. Methodologies can be classified as either "indirect" or "direct" methods of detection (Figure 4.3). Indirect detection involves the transesterification, or conversion, of the individual fatty acid esters into their corresponding free forms: either 2-MCPD, 3-MCPD, or

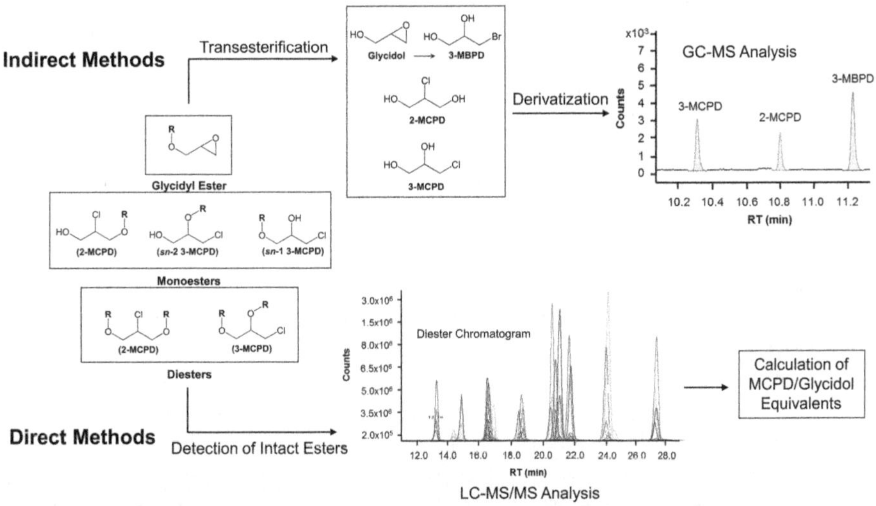

Figure 4.3 Principles of indirect and direct analysis for detection of fatty acid esters of 2-MCPD, 3-MCPD, and glycidol.

glycidol (which is typically converted into free 3-monobromopropanediol (3-MBPD), in order to avoid conversion of glycidol to 3-MCPD during sample preparation). Following transesterification, 2-MCPD, 3-MCPD, and 3-MBPD are derivatized and quantified by gas chromatography–mass spectrometry (GC-MS). Alternatively, direct detection methods involve the analysis of the individual intact esters. Oil or fat samples are typically purified (to remove components that can negatively impact chromatographic separation and/or ionization of the compounds of interest) and then analyzed by liquid chromatography–tandem mass spectrometry (LC-MS/MS) techniques. Concentrations of each individual ester are then mathematically converted into free 2-MCPD, 3-MCPD, or glycidol equivalents, since toxicity primarily stems from the free forms of the fatty acid ester contaminants.

In comparison to the direct methods, the indirect methods offer several advantages. For example, indirect techniques require just three analyte standards (and the corresponding stable, isotopically labeled standards for internal standard calibration) for the quantitation of bound 2-MCPD, 3-MCPD, and glycidol, since the individual esters present in the oil/fat are hydrolyzed during sample preparation. Conversely, direct methods involve quantitation of the intact esters, so a unique standard is required for the analysis of each compound of interest. Indirect detection can also be advantageous from a financial standpoint since analysis is carried out using GC-MS instrumentation, which is relatively inexpensive compared with LC-MS/MS techniques required for direct detection. However, there are some advantages to using direct detection. For example, direct methods allow quantitative data to be obtained for the individual intact esters present in a sample, which could be important given that the toxicity of the esters is largely unknown. Additionally, direct methodologies do not involve chemical transformations during sample preparation, eliminating the possibility of destruction or interconversion of the compounds of interest. In consideration of these points, indirect methods have been typically employed for routine analysis, particularly in regulatory and industrial settings, while direct techniques tend to be better suited for research purposes. The advantages and disadvantages of these differing detection techniques are summarized in Table 4.1.

4.4.1 Indirect Methods

The first methods developed for MCPD and glycidyl esters detection in oils and fats were exclusively indirect techniques. Typical sample preparation involves transesterification of the esters under acidic or alkaline conditions, followed by neutralization of the sample and addition of salt [sodium chloride (NaCl), sodium bromide (NaBr), or sulfate salts] to facilitate the extraction of undesirable lipophilic components in the sample. The "salting out" step is followed by the addition of a derivatization agent [typically phenylboronic acid (PBA)], which reacts with the analytes of interest to form a more volatile MCPD (or MBPD) derivative, thereby increasing method sensitivity. Samples are then analyzed using GC-MS.

Table 4.1 Advantages and disadvantages of indirect and direct detection techniques.

	Advantages	Disadvantages
Indirect methods	• Requires only three analyte standards (and corresponding isotopically labeled internal standards) for analysis • Cheaper instrumentation (GC–MS) compared with direct methods	• Some transesterification conditions may cause destruction/interconversion of compounds of interest • Some methods are time consuming (transesterification could take up to 16 h to complete)
Direct methods	• Obtain quantitative data for individual esters • Sample preparation conditions eliminate possibility of destruction or interconversion of analytes	• Considerable number of standards required for analysis (some commercially unavailable) • Expensive instrumentation (LC-MS/MS) compared with indirect methods

In 2004, Divinová *et al.* developed the first indirect method using acidic transesterification for the quantitation of bound 3-MCPD.[16] Although the method was time consuming, with hydrolysis of the esters taking approximately 16 h to complete, a number of additional methods were developed based on this novel work,[13,14,75] including methodology developed by Ermacora and Hrncirik that involves quantitation of bound 2-MCPD, 3-MCPD, and glycidol[21] that has since become an officially accepted method of the American Oil Chemists' Society (AOCS Cd 29a-13)[76] and the International Organization for Standardization (ISO 18363-3).[77] Alkaline transesterification methods were later developed, which reduced transesterification times to <10 min. However, a few of these methods were either non-specific {3-MCPD and glycidol were quantified together [Deutsche Gesellschaft fur Fettwissenschaft (DGF) C-VI 17(10)]}[78] or the use of NaCl during salting out resulted in the unwanted conversion of glycidol into 3-MCPD, resulting in overestimation of 3-MCPD concentrations. This overestimation was circumvented with the use of NaBr during salting out and led to the development of the "differential" method for quantitation of 3-MCPD and glycidol. In this method, two concurrent assays are carried out – the first using NaCl for salting out, which converts all 3-MCPD ester and glycidyl ester content into free 3-MCPD, and the second, which uses NaBr for salting out and converts only 3-MCPD esters to 3-MCPD. The final concentrations obtained from the two assays can be subtracted to indirectly determine the concentration of bound glycidol in the sample. This differential approach has been adopted as an official method by the DGF [DGF C-VI 18(10)][79], AOCS (AOCS Cd 29c-13)[80], and ISO (ISO 18363-1).[81] Although these alkaline methods significantly reduced transesterification times, the lack of specificity, inability to quantify 2-MCPD, and significant degradation of 3-MCPD to glycidol under the alkaline conditions led to the development of a method by Kuhlmann, which still involved alkaline conditions, but at much lower temperatures (−25 °C).[20] This method nearly eliminated the degradation of 3-MCPD, but required transesterification times of approximately 16 h. This method, however, is suitable for the quantitation of bound 2-MCPD, 3-MCPD, and glycidol and has been adopted as an official method by AOCS (AOCS Cd 29b-13)[82] and ISO (ISO 18363-2).[83] More recent work in this area has focused on the development of methods for the quantitation of bound 2-MCPD, 3-MCPD, and glycidol, but with much faster transesterification times so that sample preparation can be minimized, which is highly desirable in high-throughput, routine labs or in production facilities. In 2016, Zwagerman and Overman published a method based on AOCS Cd 29c-13 that was adapted for the separate quantitation of bound glycidol (as opposed to quantitation using the "difference" method) and used automated sample preparation to increase throughput.[84] Additionally, in 2019, the same group published a high-throughput method for the analysis of bound 2-MCPD, 3-MCPD, and glycidol using alkaline transesterification, but with optimized conditions that would allow sample preparation to be carried out in under 80 min.[85] This method is currently undergoing validation for AOCS and ISO approval.

4.4.2 Direct Methods

The first direct methods were developed in response to the potential unreliability of some of the indirect methods (which resulted in the conversion of glycidol into 3-MCPD under certain conditions). Therefore, it was desirable to develop methods that did not involve chemical transformations during sample preparation in order to avoid any unwanted destruction/conversion of the analytes of interest. The earliest direct methods were developed in 2010 and 2011 and involved quantitation of a limited number of esters (mainly due to the difficulty in synthesizing suitable standards or obtaining them commercially), and many of these methods suffered from a number of issues such as high limits of quantitation, instrument deterioration (due to inadequate sample purification), and poor chromatographic separation of ester isomers.[30,86–89] It was also determined that stable isotopically labeled internal standards were necessary for accurate individual ester quantitation for some analytes to avoid severe matrix effects. At present, there are still no official direct methods of analysis.

4.4.3 Methodology Developed at the U.S. FDA – Vegetable Oils

In 2011, in response to the inaccuracies of some of the indirect techniques as well as the limitations of many of the direct methodologies available, researchers at the U.S. Food and Drug Administration (FDA) began work on development of a direct methodology for the quantitation of MCPD and glycidyl esters in edible fats and oils.[22,23,25] Direct detection of the intact esters was preferred in order to address the concerns regarding the potential for undesirable chemical transformations during the sample preparation required for indirect detection. However, it was also necessary to resolve many of the challenges that were encountered during the development of the earliest direct techniques. First, it was essential that the method include a sufficient number of MCPD and glycidyl ester standards that would adequately cover the fatty acid esters present in edible oils. Too few standards would result in an underestimation of contaminant concentrations, thereby rendering the method impractical for producing meaningful occurrence data. Secondly, adequate purification of oil and fat samples would be necessary, not only for improving method accuracy and sensitivity, but to prevent instrument deterioration as well. Finally, LC conditions would need to be established such that sufficient chromatographic separation of the 2- and 3-MCPD ester isomers could be achieved, and selection of appropriate MS/MS transitions would be essential for accurate quantitation of the individual ester components.

Method development began with the selection of fatty acid ester standards necessary for coverage of the ester contaminants that would be found in edible fats and oils. This was particularly challenging given the vast majority of these unique MCPD and glycidyl ester compounds were not commercially available in the early 2010s, which meant that synthesis of many of these standards was required. Additionally, synthesis and incorporation of the extensive number of potential MCPD and glycidol fatty acid substituents into the method was impractical. To address this obstacle, the number of MCPD and glycidyl ester standards was reduced to include only those containing the seven fatty acids (lauric, myristic, palmitic, stearic, oleic, linoleic, and linolenic acids) that are most abundant in the majority of commonly consumed vegetable oils. Limiting the ester standards to just seven different fatty acid substituents ultimately resulted in the incorporation of 58 unique analyte standards into the method [28 3-MCPD diesters, 11 2-MCPD diesters, 9 (sn-1 and sn-2) 3-MCPD monoesters, 3 2-MCPD monoesters, and 7 glycidyl esters]. In addition, it was determined that isotopically labeled analyte standards were also necessary in order to avoid inaccurate quantitation due to matrix effects. However, it was not required that each analyte have an exact matching deuterated standard for quantitation, as method development demonstrated that a few deuterated ester compounds could accurately account for matrix effects for multiple analytes (eight 3-MCPD-d5-diesters, three 3-MCPD-d5-monoesters, and seven d5-glycidyl esters).

Following the selection of standards, efforts were focused on the purification of oil and fat samples. Removal of the lipophilic components of the sample, such as mono-, di-, and tri-glycerides, was essential for minimizing interferences and matrix effects and for maintaining system performance. Due to the differing polarities of the glycerides and the MCPD/

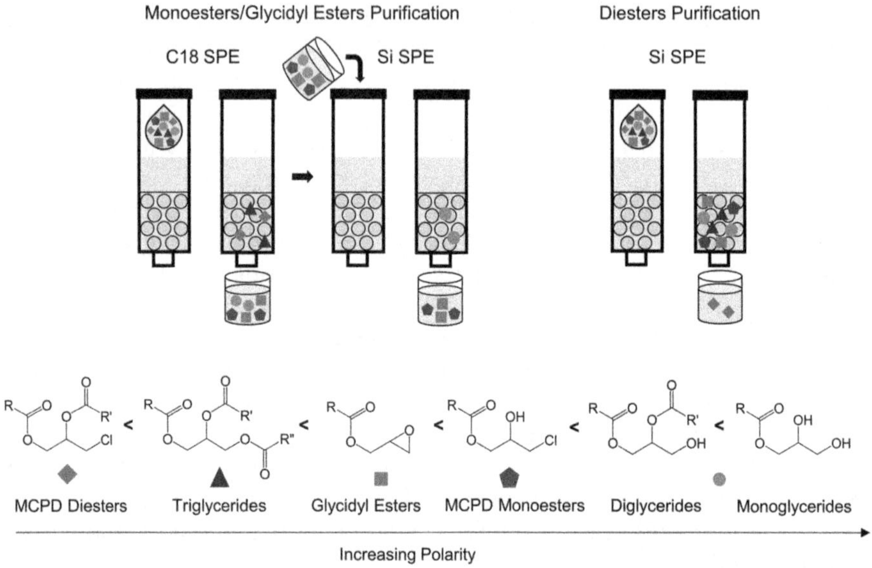

Figure 4.4 Purification of oil and fat samples according to U.S. FDA method protocol using solid-phase extraction (SPE) for the separation of MCPD monoesters/glycidyl esters and MCPD diesters.

glycidyl ester contaminants (Figure 4.4), a multi-step purification approach using solid phase extraction (SPE) was applied. In one portion of the oil or fat sample, monoesters and glycidyl esters were separated from the more non-polar components using a C18 SPE cartridge and appropriate eluent, which retained the diesters and triglycerides on the C18 stationary phase while allowing the more polar components to be eluted from the column. A second purification step, using a more polar silica (Si) SPE cartridge, allowed only the monoesters and glycidyl esters to be eluted from the column while retaining the diglycerides, monoglycerides, and any other more polar components of the sample (Figure 4.4). In a second portion of the oil or fat sample, the non-polar diesters were separated from the more polar matrix components using a Si SPE cartridge and appropriate eluent, which allowed the diesters to elute from the column while the triglycerides, glycidyl esters, monoesters, diglycerides, and monoglycerides were retained on the stationary phase (Figure 4.4).

After purification, the samples were analyzed by LC-MS/MS using separate methods for quantitation of the diesters, monoesters, and glycidyl esters in order to ensure adequate chromatographic separation of the analytes (particularly the 2- and 3-MCPD ester isomers) and enhance sensitivity. Quantitation of the monoesters and glycidyl esters was performed by monitoring the most sensitive transition, which was the loss of either the MCPD or glycidol backbone, leaving behind the positively charged fatty acid fragment. Similarly, quantitation of 2-MCPD diesters and symmetrical 3-MCPD diesters (where the two fatty acid substituents are identical) was performed by monitoring the most sensitive MS/MS transition, which involved the loss of one of the fatty acids in either the sn-2 or sn-1 position. However, these transitions were not appropriate for quantitation of the unsymmetrical 3-MCPD diesters. Because the method does not account for sn-1 and sn-2 isomerism of the 3-MCPD diesters, quantitation by monitoring the loss of one of the fatty acids would require that the ratio of the isomers in the analytical standard be equivalent to the ratio of the isomers in the unknown sample. Since this would be impossible to ensure, quantitation was performed by monitoring the loss of NH_4Cl. Although this MS/MS transition produced a fragment of relatively low sensitivity, the response of the sn-1/sn-2 3-MCPD diester isomers was identical using this transition. Single-laboratory validation results for MCPD/glycidyl ester quantitation revealed satisfactory method performance, indicating this direct technique was suitable for producing occurrence data for oils and fats in order to estimate levels of exposure.[22,23,25]

4.5 Methods of Analysis – Extension to Complex Food Matrices

Because refined oils are used as ingredients in the production of a number of foodstuffs, the presence of MCPD esters and glycidyl esters in complex food matrices has also become a point of concern. In particular, refined vegetable oils are the primary fat source in commercial infant formulas. Worldwide, infant formulas typically contain between 10 and 35% fat, and formulas manufactured and sold in the United States are almost exclusively comprised of approximately 30% refined vegetable oils. In addition to containing a substantial quantity of refined oils, formulas are also likely to be an infant's sole source of nutrition in the early stages of life. Given these factors, infant exposure to MCPD esters and glycidyl esters has become a subject of discussion in recent years and has resulted in concerted research activities to develop methods for the quantitation of these ester contaminants in infant formula and other food matrices.

The development of methods for the analysis of 3-MCPD esters and glycidyl esters in infant formula has proven to be a challenging endeavor. Extraction of these contaminants from the highly homogenized and complex mixture of fat, carbohydrates, and protein that comprise infant formulas requires a unique set of conditions that can adequately remove the entirety of the fat content from the formula, but without altering the fatty acid ester contaminants. Although initial investigations into method development have considered the use of the Rose-Gottlieb method,[90] which is traditionally used for the extraction of fat from milk powders, the harsh sample preparation conditions involving the use of ammonia has led some researchers to avoid this technique due to concerns that ammonia could react with the analytes of interest. Therefore, researchers in this field have focused their efforts on the development of effective, yet mild, fat extraction protocols, that could be paired with existing indirect or direct detection techniques for MCPD and glycidyl ester analysis.

Over the last decade, several researchers have published studies describing the extraction of the lipophilic components of infant formula and other food matrices and the subsequent analysis of MCPD and glycidyl esters in the fat extract using established indirect or direct detection methodologies. In 2009, Zelinková *et al.* described a technique that involved the extraction of fat from infant formula using a mixture of ethanol, hexane, diethyl ether, and potassium oxalate, paired with an indirect method of detection for bound 3-MCPD only.[91] However, this method was somewhat impractical for routine analysis since large volumes of organic solvent were required. A year later, Küsters *et al.* published a procedure involving the use of a sodium chloride (NaCl) solution and methyl-*t*-butyl ether (MTBE) for fat extraction, followed by indirect analysis. However, the method was not validated for infant formula products and could only be used for the quantitation of bound 3-MCPD.[19] Between 2013 and 2016 several researchers published procedures involving accelerated solvent extraction (ASE) for the removal of the fat contents from infant formulas and other processed foods.[34,92] All of the methods utilized previously validated indirect detection methods for the analysis of bound 2- and 3-MCPD, and one of the methods included bound glycidol analysis as well.[34] In 2015, researchers at Health Canada published two separate procedures for the extraction and analysis of MCPD and glycidyl esters in infant formulas.[33] Extraction of the fatty acid esters of 2- and 3-MCPD and glycidol was similar, involving the use of methanol (MeOH) and dichloromethane (CH_2Cl_2) for fat extraction, but an indirect technique was used for the analysis of bound 2- and 3-MCPD and direct detection was used for the quantitation of the glycidyl esters. Unfortunately, for the majority of these methods, because a certified reference material was not available, validation of method performance could only be evaluated using spiked samples.

Within the last year, two methods for the analysis of MCPD and glycidyl esters in infant formulas have been submitted to the Association of Official Analytical Chemists (AOAC) for consideration as official methods. One of the methods, submitted by SGS Germany, involves heated ultrasonic pressure-supported solvent extraction (HUPsSE) using mixtures of MeOH, MTBE, and isohexane, followed by indirect detection of bound MCPD and glycidol using official method AOCS 29b-13.[82] The second method, submitted by researchers from Nestlé, involves an abbreviated procedure of the extraction protocol developed by the U.S. FDA (to be discussed in subsequent sections) and indirect detection of the bound analytes using a modified version of official method AOCS 29a-13.[76,93] Both methods have been approved for collaborative study in the near future.

4.5.1 Methodology Developed at the U.S. FDA

4.5.1.1 Infant Formula

Given the concern in recent years regarding potential infant exposure to elevated concentrations of bound MCPD and bound glycidol in infant formula, researchers at the U.S. FDA began work in 2015 on the development of a method for the quantitative determination of these contaminants in formulas and other foods containing refined oils.[35] Because of the limitations of many of the previously published methods, specifically regarding the difficulty of confirming method performance, it was important to consider several critical factors during method development. First, it was essential to ensure that extraction conditions were mild enough so as not to result in the unwanted destruction/creation of the analytes of interest, but also suitable for completely extracting the target analytes of ranging polarities from the formula matrix. In addition, it was also important to consider that extraction conditions would need to be applicable to all varieties of infant formulas on the market, ranging from premium, milk-based products to specialty formulas for infants with certain medical conditions. Second, it was desirable that the fat extraction technique could be paired with the MCPD and glycidyl ester direct detection methodology developed in our laboratories at the U.S. FDA with little to no modifications to the existing method. Finally, and perhaps most importantly, it was essential to devise an approach for evaluating method performance beyond spiking the infant formula matrix with known amounts of MCPD and glycidyl esters and determining recoveries.

Initial attempts at the extraction of fat from infant formulas involved experiments using different organic solvents of various polarities [*e.g.*, heptane, MTBE, ethyl acetate (EtOAc), *etc.*] to achieve sufficient removal of the lipid contents from the matrix without inducing degradation of the analytes of interest. Efficiency of the extraction attempts was evaluated by determining the percentage of fat extracted from the infant formula based on the fat content specified on the nutrition label. Fat recoveries of ≥90% were targeted. For soy-based formulas, this fat recovery could be achieved with nearly all organic extraction solvents tested. However, fat recoveries for premium milk-based products and specialty formulations were significantly lower; due to the presence of emulsions, recoveries of only 30–80% could be reached for all solvents investigated. For this reason, a number of additional extraction techniques, such as centrifugation, heated shaking, and salt addition, were investigated to minimize emulsions and facilitate maximum fat recovery across all varieties of infant formula.

Ultimately, final extraction parameters were developed such that extracted fat yields of ≥95% were attained for all types of infant formulas investigated. As illustrated in Figure 4.5, method protocol involved vigorous shaking of infant formula samples, the extraction solvent (EtOAc), and water at 35 °C for approximately 1.5 h, followed by centrifugation at elevated

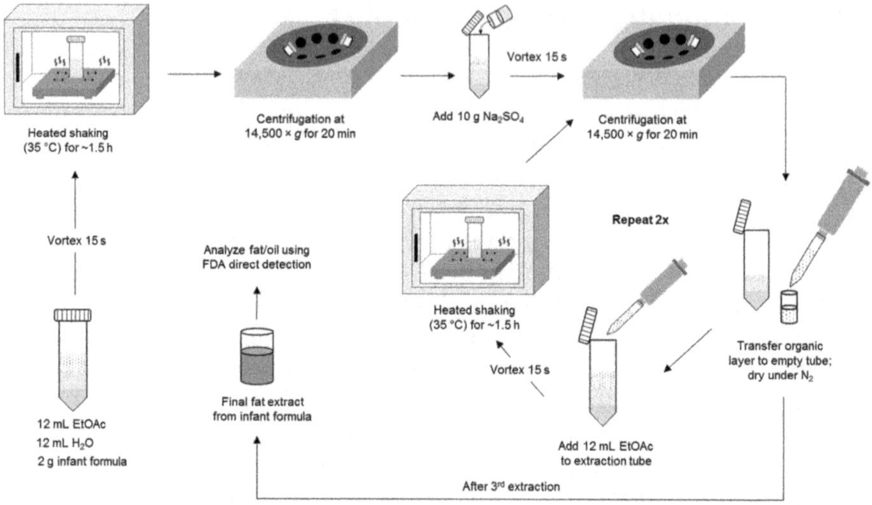

Figure 4.5 U.S. FDA method for the extraction of MCPD esters and glycidyl esters from infant formulas.

speeds (14 500×*g*). After centrifugation, emulsions were present in samples of nearly all infant formulas investigated. In order to minimize these emulsions, 10 g of sodium sulfate (Na_2SO_4) was added to the samples, followed by additional centrifugation, which thereby eliminated the emulsions and resulted in optimal extraction of the lipid contents from the formula products. Following the second centrifugation step, the organic (upper) layer of the sample was transferred to an empty vial and dried under N_2. Additional EtOAc was added to the extraction sample and the method protocol was repeated twice more. After the third and final extraction, evaporation of the combined organic solvent yielded ≥95% of the fat content from the infant formula sample, which was then analyzed for MCPD esters and glycidyl esters by the direct detection methodology previously developed in FDA's laboratories.

While the developed protocol proved effective for the extraction of fat from infant formula products, a simple determination of fat recovery was not suitable for verifying that all MCPD and glycidyl ester contaminants were adequately extracted with the lipid contents. Typical of single-laboratory validations, infant formula samples were spiked with known concentrations of MCPD esters and glycidyl esters and then extracted according to the developed protocol to determine whether the esters were indeed adequately recovered. Spike recoveries of the esters ranged between 88.7 and 107.5%, which is generally indicative of acceptable method performance. However, determination of recoveries from the spiking of standards on the surface of a solid matrix is not an accurate reflection of the actual ester extraction efficiency, since it is more difficult to extract compounds occurring within a sample. Typically, in cases such as this, certified reference materials, which have been produced with known quantities of the compounds of interest, are analyzed *via* the developed protocol in order to assess method performance. Unfortunately, a certified infant formula reference material containing known amounts of MCPD esters and glycidyl esters is not currently available. This not only made evaluation of FDA's extraction protocol challenging but has also limited the validation of many of the previously published methods.

To address this challenge of assessing method performance, researchers at the FDA developed an approach that involved following an online recipe for the preparation of a "homemade" infant formula reference material. The material was prepared with similar ingredients, including refined vegetable oils, to those typically found in commercial infant formula products. Refined oils, including palm olein, sunflower, soybean, and coconut oils, were initially analyzed for their MCPD and glycidyl ester content by FDA's direct detection methodology, and then used as ingredients in the homemade infant formula. The prepared liquid formula was then lyophilized, resulting in a powder that visually resembled commercial formula products. The homemade formula was then analyzed for its MCPD and glycidyl ester content using FDA's developed extraction protocol so that recoveries for the esters present within the powdered sample could be determined. These ester recoveries ranged between 84.9 and 109.0%, which indicated the MCPD and glycidyl esters were adequately extracted from the matrix, demonstrating suitable method performance. While it was understood that this approach did not perfectly simulate the extraction of ester contaminants from commercial products, particularly because the formulas were not produced in the same way, it was nevertheless a practical substitute in the absence of certified reference materials.

4.5.1.2 Extension to Other Complex Food Matrices

While methods have typically focused on the extraction of MCPD/glycidyl ester contaminants from infant formula due to concern of infant exposure, several exposure assessments have indicated young children may also be at risk of exposure to elevated MCPD and glycidyl ester levels, particularly through the consumption of cakes, cookies, and other bakery products containing refined oils. For this reason, it was desirable to investigate whether the FDA methodology for the analysis of ester contaminants in infant formula could also be extended to other food matrices containing refined oils. Given the complexity of the infant formula matrix, it was anticipated that FDA's extraction technique would likely be directly applicable to other food products with little to no modification of the extraction parameters. To test this hypothesis, a number of foods including snack cakes, cookies, potato chips, cheese crackers, and peanut butter were purchased from local supermarkets and their fat contents extracted according to FDA's extraction protocol for infant formula. In comparison to infant formulas, fat extraction from these products was typically less difficult, with little to no emulsions observed during sample preparation, and fat recoveries ranging between 95 and 115%. However, it was recognized that acceptable fat recoveries were not necessarily an indicator of satisfactory method performance. So, similar to the evaluation of method performance for infant formula products, it was also necessary to determine extraction recoveries for MCPD and glycidyl ester contaminants incorporated within the food matrices.

Like infant formulas, certified reference materials of food products containing known levels of MCPD esters and glycidyl esters are not currently available. Therefore, to evaluate the performance of FDA's extraction methodology for processed foods, several "homemade" food products, including cookies, cheese crackers, and peanut butter, were prepared to assess recoveries of fatty acid ester contaminants present within the food samples. A refined olive oil sample was analyzed *via* FDA's direct detection methodology to determine MCPD and glycidyl ester levels, and then incorporated into the recipes for the cookies, cheese crackers, and peanut butter. Due to concerns involving the potential destruction/creation of the analytes of interest during baking, the cookies and cheese crackers were prepared at lower than normal (≤95 °C) baking temperatures (preparation of the peanut butter did not require baking or elevated temperatures). Following preparation, the individual food products were homogenized to form powders and then analyzed according to FDA's extraction protocol for infant formula. Ester recoveries from the food products ranged from 85.8 to 114.3%, except for recoveries of glycidyl laurate (La-GE) (in peanut butter and cheese crackers) and glycidyl stearate (St-GE) (in peanut butter), which ranged between 280% and 940%. Given these exceptionally high recoveries, it was hypothesized that La-GE and St-GE may also be present in one (or more) of the ingredients (other than the refined olive oil) used to prepare the homemade food products. While this assumption requires further investigation, the elevated La-GE and St-GE recoveries (as opposed to low recoveries) combined with suitable recoveries of the other fatty acid ester contaminants were an indication that FDA's extraction protocol was effective for the analysis of MCPD and glycidyl esters in a variety of food products.

4.6 Occurrence of MCPD and Glycidyl Esters Using U.S. FDA Methodology

4.6.1 Occurrence Data – Vegetable Oils and Fats

In 2013, researchers at the FDA published an occurrence study detailing the 3-MCPD and glycidyl ester contents of more than 100 fats and oils obtained from around the world.[24] Table 4.2 displays a selection of the results from the study for 20 types of products, which represent some of the most commonly consumed oils and fats that are often present in infant formulas and other processed foods. The results were obtained using FDA's

Table 4.2 Bound 3-MCPD and bound glycidol concentrations in a selection of edible oils analyzed in an occurrence study by the U.S. FDA. Adapted from ref. 21 with permission from John Wiley and Sons, Copyright © 2013 American Oil Chemists' Society (AOCS).

Oil	No. of Samples	Bound 3-MCPD[a] (mg kg^{-1}) Range, average	Bound glycidol[b] (mg kg^{-1}) Range, average
Butter (unrefined)	4	<LOQ–0.045, 0.011	<LOQ–0.028, 0.013
Canola	7	<LOQ–0.33, 0.11	0.18–0.48, 0.28
Coconut (unrefined)	2	<LOQ	<LOQ
Coconut	7	0.025–0.38, 0.17	0.034–1.71, 0.80
Corn	9	0.06–0.42, 0.17	0.15–1.57, 0.68
Olive (unrefined)	5	<LOQ–0.025, 0.005	<LOQ
Olive	5	0.15–0.73, 0.56	0.048–1.10, 0.48
Palm (unrefined)	2	<LOQ	<LOQ
Palm	14	1.51–7.23, 3.19	0.33–10.52, 3.50
Palm kernel	2	0.038–0.20, 0.12	0.38–0.65, 0.52
Palm olein	5	1.40–8.43, 4.14	1.88–9.53, 6.03
Peanut (unrefined)	2	<LOQ	<LOQ
Peanut	3	0.14–0.69, 0.49	0.44–0.57, 0.49
Pumpkin seed	1	<LOQ	<LOQ
Safflower	5	0.28–1.77, 0.84	0.065–0.44, 0.25
Sesame (unrefined)	3	0.16–0.45, 0.29	<LOQ–0.22, 0.077
Shortening	5	0.35–0.46, 0.40	0.12–1.03, 0.46
Soybean	6	0.041–0.25, 0.11	0.014–0.50, 0.25
Sunflower (unrefined)	1	<LOQ	<LOQ
Sunflower	4	0.19–0.93, 0.55	0.012–0.90, 0.39

[a] Bound 3-MCPD LOQ: 0.002–0.06 mg kg^{-1}.
[b] Bound glycidol LOQ: 0.004–0.018 mg kg^{-1}.

direct detection protocol[22,23] and were comparable to a previous occurrence study performed by other researchers using a collaboratively studied indirect detection method.[20] The highest bound 3-MCPD and bound glycidol concentrations were observed in refined palm oils and palm-derived oils, which was consistent with previous findings. Average bound 3-MCPD concentrations for palm oil and palm olein were 3.19 and 4.14 mg kg^{-1}, respectively, and average bound glycidol concentrations for these oils were 3.50 and 6.03 mg kg^{-1}, respectively. All other oils and fats investigated contained average bound 3-MCPD levels <0.85 mg kg^{-1} (except for average concentrations of almond oil, grape seed oil, and palm stearin that are not listed in Table 4.2). Similarly, average bound glycidol contents were <0.80 mg kg^{-1} for all other fats and oils analyzed (except for average concentrations of grape seed oil and palm stearin that are not listed in Table 4.2). As expected, lowest bound 3-MCPD and glycidol concentrations were observed in unrefined oils, which is consistent with the understanding that these contaminants arise from the high-temperature deodorization step of the refining process. For nearly all unrefined fats and oils investigated, bound MCPD and glycidol levels were below the limit of quantitation (LOQ) except for unrefined sesame oils. It was determined that for each of the three oils analyzed, the sesame seeds were toasted prior to oil extraction. The use of higher temperatures for toasting could explain the elevated contaminant levels in comparison to the other unrefined oils. To date, the results of this study represent the most extensive set of occurrence data for oils and fats obtained using direct detection. In addition, the data obtained in this study, as well as occurrence data produced by other researchers, was crucial in bringing to light the potential for elevated exposure to these contaminants from the consumption of infant formulas and other processed foods containing these refined oils.

4.6.2 Occurrence Data – Infant Formula

Following development of a method for the analysis of MCPD and glycidyl esters in infant formula, researchers at the FDA published a study in 2017 detailing the occurrence of these contaminants in nearly 100 formula products purchased in the United States between 2013 and 2016.[94] Table 4.3 displays the results of this study, which lists concentration ranges and averages for formulas produced by seven different infant formula manufacturers. Based on previously published occurrence data for oils and fats,[20,24] it was not surprising that the highest 3-MCPD levels were observed in products produced by manufacturers that use palm oil or palm olein in their formulations. Formulas produced by manufacturers B, D, and F contained average concentrations of bound 3-MCPD ranging from 0.48 to 0.62 mg kg^{-1} of powdered formula. Similarly, bound glycidol levels were highest in formulas produced by manufacturers B and F, with average concentrations of 0.36 and 0.13 mg kg^{-1} powdered formula, respectively. As expected, bound 3-MCPD concentrations were lower in formulas produced by manufacturers A and G, who do not use palm oil and/or palm olein in their formulations. However, it was surprising that the lowest bound 3-MCPD levels (average 0.055 mg kg^{-1}) were observed in products produced by manufacturer E, who does use palm

Table 4.3 Bound 3-MCPD and bound glycidol concentrations in infant formulas purchased in the U.S. between 2013 and 2016. Adapted from ref. 94 with permission from Taylor & Francis Ltd (http://www.tandfonline.com), Copyright 2017.

Manufacturer	No. of Samples	Bound 3-MCPD Range, average (mg kg^{-1} powdered formula)	Bound Glycidol Range, average (mg kg^{-1} powdered formula)
A[a]	23	0.072–0.16, 0.13	0.005–0.15, 0.028
B	2	0.53–0.64, 0.59	0.33–0.40, 0.36
C	5	0.14–0.48, 0.26	0.051–0.17, 0.10
D	25	0.043–0.92, 0.48	<LOQ–0.26, 0.093
E	11	0.021–0.11, 0.055	0.011–0.057, 0.032
F	30	0.31–0.89, 0.62	0.022–0.24, 0.13
G[a]	2	0.077–0.11, 0.093	0.006–0.009, 0.008

[a]Formulas produced by this manufacturer do not contain palm olein.

olein in its infant formulas. Bound glycidol concentrations (average 0.032 mg kg^{-1}) were also remarkably low in formulas produced by manufacturer E as well. These findings were an indication that manufacturer E may have sourced refined oils that had undergone mitigation to minimize contaminant levels, or the manufacturer had implemented strategies for reducing these contaminants during the production process. In addition to these findings, it was also observed that bound glycidol concentrations were consistently lower than bound 3-MCPD concentrations in all but five samples investigated. Given that previously published occurrence data for oils and fats have shown that 3-MCPD and glycidol levels vary, with bound glycidol contents greater than bound 3-MCPD contents in many cases,[24] it is possible that conditions during infant formula production may have led to the destruction or conversion of glycidyl esters.

The FDA's occurrence data was also analyzed based on a comparison of bound 3-MCPD and glycidol concentrations in identical products purchased in 2013 and again in 2015/2016.[94] The quantitative results for a total of 21 unique products (42 individual infant formulas) produced by six different manufacturers are displayed in Table 4.4. Overall, bound 3-MCPD and glycidol levels were relatively unchanged in identical formulas purchased during the 3 year period, except for bound 3-MCPD concentrations in formulas produced by manufacturer C and bound glycidol concentrations in formulas produced by manufacturer D. From 2013 to 2016, average bound 3-MCPD concentrations decreased approximately 2-fold (0.39–0.18 mg kg^{-1} powdered formula) among manufacturer C formulas. While this could be an indication that manufacturer C began sourcing mitigated oils during this time

Table 4.4 Bound 3-MCPD and bound glycidol concentrations in identical infant formulas purchased in the U.S. in 2013 and 2015/2016. Adapted from ref. 94 with permission from Taylor & Francis Ltd (http://www.tandfonline.com), Copyright 2017.

		Year purchased: 2013 Range, average		Year purchased: 2015-2016 Range, average	
Manufacturer	No. of Samples	Bound 3-MCPD	Bound glycidol	Bound 3-MCPD	Bound glycidol
		(mg kg^{-1} powder formula)			
A	2	0.1 0–0.14, 0.12	0.011–0.017, 0.014	0.12–0.16, 0.14	0.010
C	2	0.30–0.48, 0.39	0.073–0.17, 0.12	0.14–0.23, 0.18	0.051–0.16, 0.11
D	4	0.12–0.57, 0.38	<LOQ–0.083, 0.030	0.097–0.62, 0.41	0.006–0.25, 0.094
E	1	0.076	0.047	0.079	0.032
F	11	0.35–0.84, 0.61	0.022–0.24, 0.11	0.43–0.87, 0.62	0.053–0.22, 0.14
G	1	0.11	0.006	0.077	0.009

Table 4.5 Bound 3-MCPD and bound glycidol concentrations in infant formulas purchased in Germany in 2015.

Manufacturer	No. of Samples	Bound 3-MCPD Range, average (mg kg^{-1} powdered formula)	Bound glycidol Range, average (mg kg^{-1} powdered formula)
H	9	0.029–0.18, 0.098	0.001–0.029, 0.009
I	1	0.061	0.007
J	10	0.025–0.34, 0.19	0.006–0.061, 0.025
K	8	0.013–0.067, 0.034	<LOQ–0.009, 0.003
L	3	0.073–0.29, 0.16	0.020–0.11, 0.050
M	2	0.018–0.071, 0.045	0.001–0.003, 0.002
N	17	0.046–0.12, 0.080	<LOQ–0.012, 0.003
O	3	0.019–0.024, 0.022	0.001–0.003, 0.002
P	6	0.023–0.11, 0.061	0.001–0.006, 0.004

period, this scenario is somewhat unlikely since bound glycidol levels were unaffected. In addition, only two unique sample sets produced by this manufacturer were investigated, making it impossible to draw any statistical conclusions regarding potential mitigation of contaminants. Conversely, bound glycidol concentrations increased more than 3-fold in products produced by manufacturer D (0.030–0.094 mg kg^{-1}) over the 3 year time span, which was a potential indication that the manufacturer may have changed oil suppliers during this time period. In any case, the general results taken as a whole seemed to indicate that between 2013 and 2016 U.S. manufacturers had not implemented any new or additional strategies for the reduction of MCPD and glycidol contaminants in infant formulas.

In 2015, FDA researchers collaborated with a colleague at the Technical University of Munich in Germany to investigate the occurrence of 3-MCPD and glycidyl esters in infant formulas on the German market. Fifty-nine powdered formula products produced by nine different manufacturers were purchased in 2015 and analyzed for their bound MCPD and glycidol contents (Table 4.5). Similar to data obtained for U.S. infant formulas, it was observed that bound glycidol concentrations in German formulas were always lower than bound 3-MCPD levels, which further supports the idea that infant formula production may alter contaminant contents. It was also shown that formulas produced by manufacturers J and L contained the highest average bound 3-MCPD (0.19 and 0.16 mg kg^{-1} powdered formula, respectively) and bound glycidol (0.025 and 0.050 mg kg^{-1} powdered formula, respectively) levels. However, it was noted that the highest concentrations of bound 3-MCPD and glycidol in German formulas were comparable to some of the lower levels of contaminants observed in U.S. products collected during the same time period (Table 4.3). In addition, an overall comparison of contaminant levels in German and U.S. infant formulas produced between 2013 and 2016 revealed that average bound 3-MCPD and glycidol concentrations in U.S. formulas were over four times greater and nine times greater, respectively, than contaminant levels in German formula products (Table 4.6). These results

Table 4.6 Bound 3-MCPD and bound glycidol concentrations across U.S. and german infant formulas investigated in the occurrence studies by the U.S. FDA.

Country	Bound 3-MCPD Range, average (mg kg^{-1} powdered formula)	Bound glycidol Range, average (mg kg^{-1} powdered formula)
United States (98 formulas, 2013–2016)	0.021–0.92, 0.38	<LOQ–0.40, 0.089
Germany (59 formulas, 2015)	0.013–0.34, 0.094	<LOQ–0.081, 0.010

Table 4.7 Bound 3-MCPD and glycidol concentrations in processed food products containing refined oils.

Food[a]	No. of Samples	Bound 3-MCPD Range, average (mg kg^{-1} food sample)	Bound glycidol Range, average (mg kg^{-1} food sample)
Chips (with palm)	2	0.010–0.042, 0.026	0.080–0.20, 0.14
Chips (no palm)	5	0.015–0.81, 0.20	0.021–0.16, 0.095
Cookies (with palm)	17	0.001–0.68, 0.13	0.009–0.46, 0.14
Cookies (no palm)	0	NA[b]	NA[b]
Snack cakes (with palm)	8	0.058–0.21, 0.14	0.028–0.19, 0.092
Snack cakes (no palm)	1	0.006	0.005

[a]Foods contain 10–36% fat.
[b]NA, not applicable.

were a strong indication that, during this time period, manufacturers of German infant formulas were likely sourcing refined oils for their products that had undergone mitigation to minimize MCPD and glycidyl esters levels, and that these manufacturers as a whole were more advanced in the reduction of contaminant levels in their products than their U.S. counterparts. However, preliminary results of a recent study conducted by FDA researchers regarding contaminant levels in formula products produced by U.S. manufacturers in 2018 revealed that bound 3-MCPD and glycidol concentrations were significantly reduced in comparison to data collected between 2013 and 2016. Full study results are currently being evaluated.

4.6.3 Occurrence Data – Other Complex Foods

In addition to infant formula products, researchers at the FDA also investigated the occurrence of MCPD and glycidyl esters in several processed food products containing refined oils and fats. Various varieties of chips, cookies, and snack cakes, totaling 38 unique products, were analyzed. As listed in Table 4.7, bound 3-MCPD concentrations in the various food products were generally similar to or slightly lower than bound 3-MCPD levels in the U.S. infant formulas with the lowest observed contaminant levels from the

2013–2016 occurrence study (Table 4.3). However, the bound glycidol concentrations in these food products were more on par with the highest levels of bound glycidol contaminants in the same U.S. infant formulas. In addition, in contrast to the results of the 2013–2016 infant formula occurrence study, bound glycidol levels were not always lower than bound 3-MCPD concentrations, which is more consistent with previously published occurrence data in refined oils and fats.

Evaluation of the individual food products based on their palm oil content revealed that various varieties of chips containing palm oil had significantly lower average bound 3-MCPD levels (0.026 mg kg^{-1} food) than chips that did not contain palm oil (0.20 mg kg^{-1} food), which was somewhat unexpected since previously published work has indicated palm and palm-derived oils generally contain the highest levels of ester contaminants.[20,24] However, upon further inspection of the individual data for each chips product, it was discovered that only one of the products without palm oil contained exceptionally high levels of bound 3-MCPD (0.81 mg kg^{-1} food). Removal of this "outlier" from the "chips (no palm)" category resulted in an average bound 3-MCPD concentration of 0.036 mg kg^{-1} food, which was then relatively similar to the average bound 3-MCPD level for "chips (with palm)". Average bound glycidol levels for both categories of chip were generally similar (0.14 and 0.095 mg kg^{-1} food); removing the "outlier" product containing a high concentration of bound 3-MCPD from the "chips (no palm)" category reduced the average bound glycidol concentration to 0.08 mg kg^{-1} food product.

All cookies investigated contained palm oil, with average contaminant concentrations similar to those observed for snack cakes containing palm oil. Contaminant levels in the snack cake product that did not contain palm oil were near the LOQ; evaluation of the ingredients revealed the majority of the fat content in this product was cottonseed oil, which, according to previously published occurrence studies, generally contains lower levels of ester contaminants. This product also contained just 10.4% fat, which would result in lower contaminant levels per gram of food product. Unfortunately, given the sample size, it was not possible to draw any further conclusions regarding the potential use of mitigated oils in the food products investigated. Differing manufacturers, varieties of products, food processing conditions, and fat contents could all impact contaminant levels, which made it difficult to truly compare results. Further studies, requiring a much larger sample size, are required.

4.7 Conclusions and Future Perspectives

Over the last 15 years, the body of research encompassing the analysis, occurrence, and mitigation of MCPD and glycidyl esters in refined edible oils/fats and complex food matrices containing these oils has expanded tremendously. At present, numerous analytical methods for the quantitation of these ester contaminants in refined oils have been published, including three collaboratively studied procedures that have been adopted as official methods by several professional organizations. While recent years have seen the development of methods for the analysis of these compounds in various food matrices, particularly infant formula, official methods have not yet been adopted. In the last few years, researchers at the U.S. FDA have published methods for both the direct detection of MCPD and glycidyl esters in refined oils as well as the extraction of these contaminants from infant formula and other processed foods. These single-laboratory validated methods were used to carry out studies on the occurrence of these contaminants in oils and infant formulas from around the world, as well as a number of processed food products purchased in the United States. Analysis of the refined oil products revealed that the highest ester concentrations were observed in palm and palm-derived oils, which was consistent with previously published studies. Ester concentrations in U.S. infant formula products purchased between 2013 and 2016 varied widely by manufacturer, and on average were substantially higher than contaminant levels in infant formula products purchased on the German market in 2015. These results suggested that manufacturers of German infant formula were likely sourcing refined oils for their products that had undergone mitigation to reduce contaminant levels, although based on ester concentrations it appeared that one U.S. manufacturer may be sourcing mitigated oils as well. The investigation of MCPD and glycidyl ester concentrations in various food products revealed average contaminant levels that were slightly lower than those observed in U.S. infant formula, but due to a multitude of varying factors such as differing manufacturers, food varieties, and fat content it was not possible to draw any further conclusions from the small sample set.

The recent implementation of EU guidelines for maximum allowable concentrations of bound glycidol in refined oils/fats and infant formulas has not only highlighted the need for accurate and robust methods of analysis, but also for mitigation strategies that effectively reduce ester contaminants in these products so that they remain in compliance with these regulations. Future work regarding analysis of these ester contaminants in infant formulas and other processed foods will likely focus on the acceptance of "official" methods of analysis. Collaborative studies of these techniques will be necessary to ensure accuracy and robustness of the methods, but could be challenging since certified reference materials are not currently available. Future research efforts may focus on the development of such materials or determining other adequate means of ensuring method performance. Modifications of current techniques or the development of new analytical methodologies may also be necessary if routine analysis requires high-throughput. In addition, very little research is available regarding the impact of production methods on MCPD and glycidyl ester levels in infant formula and processed foods and will therefore likely be a topic of future investigations. Research on the mitigation of these contaminants is still a vastly open field. The coming years will likely see the emergence of innovative technologies for the reduction of these contaminants in oils and processed foods, as well as the implementation of many of the current and emerging mitigation strategies on an industrial scale.

References

1. J. Velíšek, J. Davidek, J. Hajšlová, V. Kubelka, G. Janíček and B. Mánková, *Z. Lebensm.-Unters. Forsch.*, 1978, **167**, 241.
2. J. Velíšek, J. Davidek, V. Kubelka, G. Janíček, Z. Svobodová and Z. Šimicová, *J. Agric. Food Chem.*, 1980, **28**, 1142.
3. J. Davidek, J. Velíšek, V. Kubelka, G. Janíček and Z. Šimicová, *Z. Lebensm.-Unters. Forsch.*, 1980, **171**, 14.
4. B. Q. Lee and S. M. Khor, *Compr. Rev. Food Sci. Food Saf.*, 2015, **14**, 48.
5. P. D. Collier, D. D. O. Cromie and A. P. Davies, *J. Am. Oil Chem. Soc.*, 1991, **68**, 785.
6. D. C. Meierhans, S. Bruehlmann, J. Meili and C. Taeschler, *J. Chromatogr. A*, 1998, **802**, 325.
7. R. MacArthur, C. Crews, A. Davies, P. Brereton, P. Hough and D. Harvey, *Food Addit. Contam.*, 2000, **17**, 903.
8. P. J. Nyman, G. W. Diachenko and G. A. Perfetti, *Food Addit. Contam.*, 2003, **20**, 909.
9. I. Baer, B. de la Calle and P. Taylor, *Anal. Bioanal. Chem.*, 2010, **396**, 443.
10. C. Crews, P. Hough, P. Brereton, D. Harvey, R. MacArthur and W. Matthews, *Food Addit. Contam.*, 2002, **19**, 22.
11. V. Divinová, B. Svejkovská, O. Novotný and J. Velíšek, *Czech J. Food Sci.*, 2004, **22**, 267.
12. C. G. Hamlet, P. A. Sadd, C. Crews, J. Velíšek and D. E. Baxter, *Food Addit. Contam.*, 2002, **19**, 619.
13. Z. Zelinková, B. Svejkovská, J. Velíšek and M. Doležal, *Food Addit. Contam.*, 2006, **23**, 1290.
14. W. Seefelder, N. Varga, A. Studer, G. Williamson, F. P. Scanlan and R. H. Stadler, *Food Addit. Contam., Part A*, 2008, **25**, 391.
15. R. Weißhaar, *Eur. J. Lipid Sci. Technol.*, 2011, **113**, 304.
16. V. Divinová, B. Svejkovska, M. Doležal and J. Velíšek, *Czech J. Food Sci.*, 2004, **22**, 182.
17. V. Divinová, M. Doležal and J. Velíšek, *Czech J. Food Sci.*, 2007, **25**, 39.
18. B. Svejkovská, O. Novotný, V. Divinová, Z. Reblová, M. Doležal and J. Velíšek, *Czech J. Food Sci.*, 2004, **22**, 190.
19. M. Küsters, U. Bimber, A. Ossenbrüggen, S. Reeser, R. Gallitzendörfer and M. Gerhartz, *J. Agric. Food Chem.*, 2010, **58**, 6570.
20. J. Kuhlmann, *Eur. J. Lipid Sci. Technol.*, 2011, **113**, 335.
21. A. Ermacora and K. Hrncirik, *J. Am. Oil Chem. Soc.*, 2013, **90**, 1.

22. S. MacMahon, E. Mazzola, T. H. Begley and G. W. Diachenko, *J. Agric. Food Chem.*, 2013, **61**, 4737.
23. S. MacMahon, T. H. Begley and G. W. Diachenko, *J. Agric. Food Chem.*, 2013, **61**, 4748.
24. S. MacMahon, T. H. Begley and G. W. Diachenko, *Food Addit. Contam., Part A*, 2013, **30**, 2081.
25. S. MacMahon, C. D. Ridge and T. H. Begley, *J. Agric. Food Chem.*, 2014, **62**, 11647.
26. J. Kuhlmann and R. Weißhaar, 2nd Workshop on Analysis of 3-MCPD-Esters in Edible Oils, Federal Institute for Risk Assessment (BfR), Berlin, Germany, November 2008, URL: http://www.researchgate.net/publication/327890441.
27. M. W. Collison, Association of the Oilseed Crushing and Oil Refining Industry of Germany (OVID), Berlin, Germany, January 25, 2010, URL: http://www.ovidverband.de/fileadmin/user_upload/ovid-verband.de/downloads/ADM_Collison.pdf.
28. R. Weißhaar and R. Perz, *Eur. J. Lipid Sci. Technol.*, 2010, **112**, 158.
29. H. Shiro, N. Kondo, N. Kibune and Y. Masukawa, *Eur. J. Lipid Sci. Technol.*, 2011, **113**, 356.
30. T. D. Haines, K. J. Adlaf, R. M. Pierceall, I. Lee, P. Venkitasubramanian and M. W. Collison, *J. Am. Oil Chem. Soc.*, 2011, **88**, 1.
31. M. R. Blumhorst, M. W. Collison, R. Cantrill, H. Shiro, Y. Masukawa, S. Kawai and K. Yasunaga, *J. Am. Oil Chem. Soc.*, 2013, **90**, 493.
32. H. Steenbergen, K. Hrnčiřík, A. Ermacora, S. de Koning and H.-G. Janssen, *J. Chromatogr. A*, 2013, **1313**, 202.
33. A. Becalski, T. Zhao, S. Feng and B. P. Y. Lau, *J. Food Compos. Anal.*, 2015, **44**, 111.
34. F. Wöhrlin, H. Fry, M. Lahrssen-Wiederholt and A. Preiß-Weigert, *Food Addit. Contam., Part A*, 2015, **32**, 1810.
35. J. K. Leigh and S. MacMahon, *J. Agric. Food Chem.*, 2016, **64**, 9442.
36. A. Ermacora and K. Hrnčiřík, *Food Addit. Contam., Part A*, 2014, **31**, 985.
37. European Commission (EC), *Commission Regulation 2018/290*, Brussels, Belgium, 2018.
38. EFSA Panel on Contaminants in the Food Chain (CONTAM), *EFSA J.*, 2018, **16**, 1583.
39. N. Bakhiya, K. Abraham, R. Gürtler, K. E. Appel and A. Lampen, *Mol. Nutr. Food Res.*, 2011, **55**, 509.
40. B. Schilter, G. Scholz and W. Seefelder, *Eur. J. Lipid Sci. Technol.*, 2011, **113**, 309.
41. T. Buhrke, R. Weißhaar and A. Lampen, *Arch. Toxicol.*, 2011, **85**, 1201.
42. T. Buhrke, F. Frenzel, J. Kuhlmann and A. Lampen, *Arch. Toxicol.*, 2015, **89**, 2243.
43. K. Abraham, K. E. Appel, E. Berger-Preiss, E. Apel, S. Gerling, H. Mielke, O. Creutzenberg and A. Lampen, *Arch. Toxicol.*, 2013, **87**, 649.

44. K. E. Appel, K. Abraham, E. Berger-Preiss, T. Hansen, E. Apel, S. Schuchardt, C. Vogt, N. Bakhiya, O. Creutzenberg and A. Lampen, *Arch. Toxicol.*, 2013, **87**, 1649.
45. International Agency for Research on Cancer (IARC), *IARC Monographs on Some Chemicals Present in Industrial and Consumer Products, Food and Drinking-water: 3-Monochloro-1,2-propanediol*, Lyon, France, 2016, vol. 101, pp. 349–374.
46. E. Barocelli, A. Corradi, A. Mutti and P. G. Petronini, Comparison between 3-MCPD and its palmitic esters in a 90-day toxicological study: Scientific Report Submitted to EFSA, 2011. URL: https://efsa.onlinelibrary.wiley.com/doi/epdf/10.2903/sp.efsa.2011.EN-187.
47. W.-S. Cho, B. S. Han, K. T. Nam, K. Park, M. Choi, S. H. Kim, J. Jeong and D. D. Jang, *Food Chem. Toxicol.*, 2008, **46**, 3172.
48. Joint FAO/WHO Expert Committee on Food Additives (JECFA), WHO Technical Report Series: Eighty-Third Report of *Evaluation of certain contaminants in food: 3-MCPD Esters and 3-MCPD*, Geneva, Switzerland, 2017, pp. 90–106.
49. International Agency for Research on Cancer (IARC), *IARC Monographs on Some Industrial Chemicals: Glycidol*, Lyon, France, 2000, vol. 77, pp. 469–486.
50. National Toxicology Program (NTP), U.S. Department of Health and Human Services, NTP Technical Report on the Toxicology and Carcinogenesis Studies of Glycidol in F344/N Rats and B6C3F1 Mice, Research Triangle Park, NC, 1990, Technical Report Series No. 374.
51. H. Akane, A. Shiraki, N. Imatanaka, Y. Akahori, M. Itahashi, H. Abe and M. Shibutani, *Toxicol. Lett.*, 2014, **224**, 424.
52. Food Safety Commission of Japan (FSCJ), Considerations on glycidol and its fatty acid esters in foods, Risk assessment report-Novel Foods and Food Additives FS/185/2015, Japan, 2015, URL: http://www.fsc.go.jp/english/evaluationreports/others/annex_glycidol_26-52.pdf.
53. EFSA Panel on Contaminants in the Food Chain (CONTAM), *EFSA J.*, 2016, **14**, 1.
54. M. Liu, B.-Y. Gao, F. Qin, P.-P. Wu, H.-M. Shi, W. Luo, A.-N. Ma, Y.-R. Jiang, X.-B. Xu and L.-L. Yu, *Food Chem. Toxicol.*, 2012, **50**, 3785.
55. M. Liu, J. Liu, Y. Wu, B. Gao, P. Wu, H. Shi, X. Sun, H. Huang, T. T. Wang and L. Yu, *J. Sci. Food Agric.*, 2017, **97**, 841.
56. G. F. N. Ebongue, P. Koona, B. Nouy, S. Zok, F. Carrière, P.-H. A. Zollo and V. Arondel, *Eur. J. Lipid Sci. Technol.*, 2008, **110**, 505.
57. F. Destaillats, B. D. Craft, L. Sandoz and K. Nagy, *Food Addit. Contam., Part A*, 2012, **29**, 29.
58. B. D. Craft, K. Nagy, L. Sandoz and F. Destaillats, *Food Addit. Contam., Part A*, 2012, **29**, 354.
59. FAO/WHO Codex Alimentarius Commission, *Appendix VI: Proposed Draft Code of Practice for the Reduction of 3-Monochloropropane-1,2-diol Esters (3-MCPDE) and Glycidyl Esters (GE) in Refined Oils and Food Products Made*

with Refined Oils, Utrecht, The Netherlands, 2018, URL: http://www.fao.org/fao-who-codexalimentarius/sh-proxy/en/?lnk=1&url=https%253A%252F%252Fworkspace.fao.org%252Fsites%252Fcodex%252FMeetings%252FCX-735-12%252FREPORT%252520%2528FINAL%2529%252FREP18_CFe.pdf.
60. K. Nagy, L. Sandoz, B. D. Craft and F. Destaillats, *Food Addit. Contam., Part A*, 2011, **28**, 1492.
61. F. Pudel, P. Benecke, P. Fehling, A. Freudenstein, B. Matthäus and A. Schwaf, *Eur. J. Lipid Sci. Technol.*, 2011, **113**, 368.
62. B. Matthäus and F. Pudel, *Lipid Technol.*, 2013, **25**, 151.
63. M. Ramli, S. Lin, N. Ibrahim, R. Hussein, A. Kuntom, R. A. Abd Razak and K. Nesaretnam, *J. Am. Oil Chem. Soc.*, 2011, **88**, 1839.
64. M. Zulkurnain, O. M. Lai, R. A. Latip, I. A. Nehdi, T. C. Ling and C. P. Tan, *Food Chem.*, 2012, **135**, 799.
65. A. Freudenstein, J. Weking and B. Matthäus, *Eur. J. Lipid Sci. Technol.*, 2013, **115**, 286.
66. B. D. Craft, K. Nagy, W. Seefelder, M. Dubois and F. Destaillats, *Food Chem.*, 2012, **132**, 73.
67. K. Hrncirik and G. van Duijn, *Eur. J. Lipid Sci. Technol.*, 2011, **113**, 374.
68. U. Strijowski, V. Heinz and K. Franke, *Eur. J. Lipid Sci. Technol.*, 2011, **113**, 387.
69. W. Cheng, G. Liu, X. Wang and L. Han, *J. Agric. Food Chem.*, 2017, **65**, 9753.
70. M. Shimizu, J. Moriwaki, D. Shiiba, H. Nohara, N. Kudo and Y. Katsuragi, *J. Oleo Sci.*, 2012, **61**, 23.
71. U. T. Bornscheuer and M. Hesseler, *Eur. J. Lipid Sci. Technol.*, 2010, **112**, 552.
72. B. A. Mogol, C. Pye, W. Anderson, C. Crews and V. Gökmen, *J. Agric. Food Chem.*, 2014, **62**, 7297.
73. FEDIOL, Report: *MCPD Esters and Glycidyl Esters: Review of Mitigation Measures*, Bussells, Belgium, 2015, URL: http://www.fediol.be/data/FEDIOL%20Review%20of%20Mitigation%20Measures%20MCPD%20Esters%20and%20Glycidyl%20Esters%20-%2024%20June%202015.pdf.
74. German Federation for Food Law and Food Science (BLL), Report: *Toolbox for the Mitigation of 3-MCPD Esters and Glycidyl Esters in Food*, Berlin, Germany, 2016, URL: http://www.bll.de/de/infomaterial/toolboxen/pb-toolbox-3-mcpd-flycidyl-fe.
75. A. Ermacora and K. Hrncirik, *J. Am. Oil Chem. Soc.*, 2012, **89**, 211.
76. American Oil Chemists' Society (AOCS), AOCS Official Method Cd 29a-13: 2- and 3-MCPD Fatty Acid Esters and Glycidol Fatty Acid Esters in Edible Oils and Fats by Acid Transesterification and GC/MS, in *Official Methods and Recommended Practices of the AOCS*, Urbana, IL, 7th edn, 2017.
77. International Organization for Standardization (ISO), ISO 18363-3:2017: Determination of fatty-acid-bound chloropropanediols (MCPDs) and

glycidol by GC/MS; Part 3: Method using acid transesterification and measurement for 2-MCPD, 3-MCPD and glycidol, in *ISO Standards Catalogue*, Geneva, Switzerland, 2017.

78. Deutsche Gesellschaft fur Fettwissenschaft (DGF), DGF Standard Method C-VI 17(10): Ester-bound 3-chloropropane-1,2-diol (3-MCPD ester) and glycidol (glycidyl ester). Summation method for the determination in fats and oils by GC-MS, in *DGF Standard Methods*, Stuttgart, Germany, 2nd edn, 2018.
79. Deutsche Gesellschaft fur Fettwissenschaft (DGF), DGF Standard Method C-VI 18(10): Fatty-acid-bound 3-chloropropane-1,2-diol (3-MCPD) and 2,3-epoxipropane1-ol (glycidol). Determination in oils and fats by GC/MS (Differential measurement), in *DGF Standard Methods*, Stuttgart, Germany, 2nd edn, 2018.
80. American Oil Chemists' Society (AOCS), AOCS Official Method Cd 29c-13: 2- and 3-MCPD Fatty Acid Esters and Glycidol Fatty Acid Esters in Edible Oils and Fats by GC/MS (Difference Method), in *Official Methods and Recommended Practices of the AOCS*, Urbana, IL, 7th edn, 2017.
81. International Organization for Standardization (ISO), ISO 18363–1:2015: Determination of fatty-acid-bound chloropropanediols (MCPDs) and glycidol by GC/MS; Part 1: Method using fast alkaline transesterification and measurement for 3-MCPD and differential measurement for glycidol, in *ISO Standards Catalogue*, Geneva, Switzerland, 2015.
82. American Oil Chemists' Society (AOCS), AOCS Official Method Cd 29b-13: 2- and 3-MCPD Fatty Acid Esters and Glycidol Fatty Acid Esters in Edible Oils and Fats by Alkaline Transesterification and GC/MS, in *Official Methods and Recommended Practices of the AOCS*, Urbana, IL, 7th edn, 2017.
83. International Organization for Standardization (ISO), ISO 18363–2:2018: Determination of fatty-acid-bound chloropropanediols (MCPDs) and glycidol by GC/MS; Part 2: Method using slow alkaline transesterification and measurement for 2-MCPD, 3-MCPD and glycidol, in *ISO Standards Catalogue*, Geneva, Switzerland, 2018.
84. R. Zwagerman and P. Overman, *Eur. J. Lipid Sci. Technol.*, 2016, **118**, 997.
85. R. Zwagerman and P. Overman, *Eur. J. Lipid Sci. Technol.*, 2019, **121**, 1800395.
86. Y. Masukawa, H. Shiro, N. Kondo and N. Kudo, *J. Am. Oil Chem. Soc.*, 2011, **88**, 15.
87. K. Hori, N. Koriyama, H. Omori, M. Kuriyama, T. Arishima and K. Tsumura, *Food Sci. Technol. Int.*, 2012, **48**, 204.
88. E. Moravcova, L. Vaclavik, O. Lacina, V. Hrbek, K. Riddellova and J. Hajslova, *Anal. Bioanal. Chem.*, 2012, **402**, 2871.
89. K. Yamazaki, M. Ogiso, S. Isagawa, T. Urushiyama, T. Ukena and N. Kibune, *Food Addit. Contam., Part A*, 2013, **30**, 52.

90. Food and Agriculture Organization (FAO) of the United Nations, Milk Fat (Rose-Gottlieb Method), in *Food Analysis: Quality, Adulteration and Tests of Identity*, Rome, Italy, 1997, vol. 14/8, pp. 10–11.
91. Z. Zelinková, M. Doležal and J. Velíšek, *Eur. Food Res. Technol.*, 2009, **228**, 571.
92. R. Jędrkiewicz, A. Głowacz-Różyńska, J. Gromadzka, A. Kloskowski and J. Namieśnik, *Food Anal. Methods*, 2016, **9**, 3460.
93. M. Dubois, A.-M. Empl, G. Jaudzems, Q. Basle and E. Konings, *J. AOAC Int.*, 2019, **102**, 903.
94. J. Leigh and S. MacMahon, *Food Addit. Contam., Part A*, 2017, **34**, 356.

CHAPTER 5

Strategies to Mitigate MCPD and Glycidyl Esters in Refined Oils and Foods

WILLIAN CRUZEIRO SILVA[a], ROSELI APARECIDA FERRARI[b], EDUARDO VICENTE[b], KLICIA ARAUJO SAMPAIO[a] AND ADRIANA PAVESI ARISSETO*[a]

[a]School of Food Engineering, University of Campinas (UNICAMP), Rua Monteiro Lobato 80, 13083-862 Campinas – SP, Brazil; [b]Institute of Food Technology (ITAL), Avenida Brasil 2880, C. P. 139, 13070-178 Campinas – SP, Brazil

*E-mail: adriana.arisseto@gmail.com

5.1 Introduction

The esterified forms of monochloropropanediols (MCPDE) and glycidol (GE) are a group of processing contaminants formed during edible oil processing, especially in the refining steps in which high temperatures are employed.[1] MCPDEs are formed from acylglycerols and chlorides, presenting two positional isomers known as 3-monochloropropane-1,2-diol esters (3-MCPDE) and 2-monochloropropane-1,3-diol esters (2-MCPDE). GE is an organic compound structurally characterized by a glycerol molecule containing epoxide and alcohol functional groups. These substances have attracted worldwide attention since their free forms (3-MCPD, 2-MCPD and glycidol) have shown toxicity and carcinogenicity in animal models.[2]

Food Chemistry, Function and Analysis No.19
Mitigating Contamination from Food Processing
Edited by Catherine S. Birch and Graham A. Bonwick
© The Royal Society of Chemistry 2020
Published by the Royal Society of Chemistry, www.rsc.org

In recent years, significant levels of 3-MCPDE, 2-MCPDE, and GE have been found in processed foods and edible oils. The highest concentrations of these contaminants have been observed in palm oil; however, other vegetable oils have presented substantial amounts, such as palm kernel oil, corn oil, and coconut oil, among others.[1] In addition, high concentrations have been found in foodstuffs that use refined oils in their production such as infant formulas, biscuits, crackers, and potato chips.[1,3]

These compounds are not present in the raw material, but can be formed through chemical reactions under high temperatures, such as those employed in the deodorization step of the refining process of oils. Above 140 °C, detectable amounts of these contaminants can be found in refined oils.[4] Depending on the type of oil, the deodorization step requires high temperatures (180–270 °C) to remove free fatty acids (FFA) and other compounds, which is a critical factor for the formation of these contaminants.[5–7] Other factors have been identified to play an important role in the contamination of refined oils, such as the presence of potential precursors that include chlorinated compounds and partial acylglycerols.

Considering the consumer exposure to these chemicals through the diet and potential risks to health, it is of the utmost importance to develop means to mitigate the formation of these contaminants in food products. Thus, some strategies have been investigated using different approaches that include: removal of precursors from the crude oil; modifications of the processing parameters; addition of substances that can prevent the formation of the contaminants during processing; and removal of the contaminants formed in the refined oil.[8,9] This chapter provides a comprehensive review of these strategies, as well as an overview of toxicity, formation, occurrence, regulation and methods of analysis for 3-MCPDE, 2-MCPDE and GE in refined oils and foods.

5.1.1 Chemical and Physical Properties

Figure 5.1 illustrates the chemical structures of free and bound forms of monochloropropanediols and glycidol. For MCPD, both of them occur in several foods and food ingredients, but the esterified form, bound to fatty acids, is the most common.[1] For glycidol, only the esters are found in foods.

Free 3-MCPD (molecular weight of 110.54 g mol^{-1}) is a hygroscopic colorless liquid with a pleasant odor. It is soluble in water, alcohol, diethyl ether, and acetone. Its boiling point is 213 °C at 760 mm Hg.[1] 2-MCPD is a positional isomer of 3-MCPD. Their chemical structures differ only in the position of the chlorine atom which confers on 2-MCPD similar chemical and physical properties to 3-MCPD. As 3-MCPD, 2-MCPD has a molecular weight of 110.54 g mol^{-1} and density of 1.3218 g cm^{-3} at 20 °C.[1] Glycidol is a colorless and odorless liquid with a molecular weight of

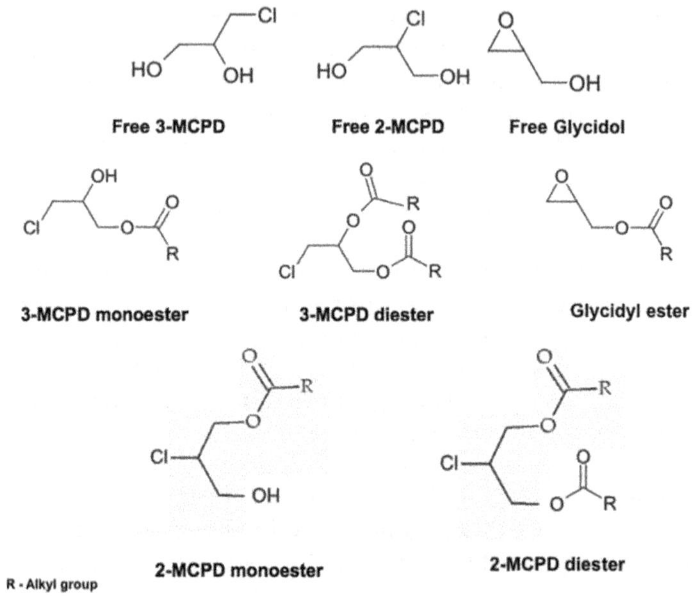

Figure 5.1 Chemical structures of the contaminants in their free and esterified forms.

74.08 g mol^{-1}. Its boiling point is 162 °C and its density at 20 °C is 1.143 g cm^{-3}. It is miscible in water, alcohols, ketones, esters, ether and aromatic compounds.[10]

Considering the bound forms, the most common fatty acids that esterify to 3-MCPD, 2-MCPD and GE are lauric acid (C12:0), myristic acid (C14:0), palmitic acid (C16:0), stearic acid (C18:0), oleic acid (C18:1), linoleic acid (C18:2) and linolenic acid (C18:3).[11] Since glycidol has only one hydroxyl in its structure, it can be esterified to one fatty acid, whereas 3-MCPD and 2-MCPD can each form monoesters and diesters. When bound to fatty acids, the solubility of these compounds increases in non-polar solvents and decreases in polar solvents given the non-polar character of fatty acids.[1]

5.2 Toxicological Aspects

Toxicity has been mainly attributed to the free forms of monochloropropanediols (3-MCPD, and 2-MCPD) and GE. However, the esterified forms are hydrolyzed by lipases in the intestinal tract releasing the free forms and eliciting potential toxicity.[12]

3-MCPD presents testicular and renal toxicity as well as the potential to induce cancer in experimental animals while glycidol is considered a genotoxic carcinogen. According to the International Agency for Research on Cancer (IARC), 3-MCPD is classified as a possible human carcinogen (group 2B) and glycidol as a probable human carcinogen (group 2A).[10,13] There is no carcinogenicity classification for 2-MCPD given the limited data available in the literature.[1] Nevertheless, it has been observed that 2-MCPD presents toxicity in rat kidneys just as 3-MCPD, but by different mechanisms.[14]

After oral intake, 3-MCPD esters are slowly absorbed in the gastrointestinal tract, with a peak concentration in plasma after 2–3 h. The slow absorption can be explained by the hydrolysis reactions of the esters in the intestinal tract that retards 3-MCPD absorption.[12] Experimental data shows that 3-MCPD diesters reach the maximum serum concentration of 135 ng mL^{-1} at about 2.5 h after oral intake of 1600 mg kg^{-1} of 3-MCPD dipalmitate. Although the 3-MCPD main target organs are kidneys and testis, its metabolites have been identified in other sites such as liver, brain, thymus, intestine, spleen, and plasma.[15] *In vivo* genotoxicity has not been observed for 3-MCPD and 3-MCPDE.[16]

A preliminary study by Huang *et al.* (2018), showed that 3-MCPDE may present immunotoxicity, since an adverse effect on T lymphocyte-mediated immunity, which involves the attenuation of MAPK (mitogen-activated protein kinase) and NF-κB (nuclear factor kappa-light-chain-enhancer of activated B cells) signaling pathways, was observed.[17]

Glycidol presents genotoxicity due to the high reactivity of the epoxide ring in its structure. The electrophile nature of the epoxide ring allows glycidol to bind covalently to DNA and hemoglobin resulting in DNA adducts such as 3-(2,3 dihydroxypropyl)-dUrd and 3-(2,3-dihydroxypropyl)-dThd.[18]

For risk assessment purposes, a tolerable daily intake (TDI) is generally used for 3-MCPD, which is an estimate of the amount that can be consumed over a lifetime without presenting an appreciable risk to health. The TDI established by the European Food Safety Authority (EFSA) is 2 μg kg^{-1} body weight (bw). Exposure data shows that the intake of 3-MCPD slightly exceeds the current TDI for some population groups, which may pose a health risk especially for high consumers in younger age groups fed with infant formula only.[19] Although the Joint FAO/WHO Expert Committee on Food Additives (JECFA) derived a different Provisional Maximum Tolerable Daily Intake (PMTDI, 4 μg kg^{-1} bw), the conclusions about the potential risks to infants were similar to those from EFSA.[20]

As glycidol is a substance that presents both genotoxicity and carcinogenicity, the margin of exposure (MOE) approach has been used for risk assessment. The MOE is a ratio between two factors which considers the dose at which a small but measurable response of an adverse effect is observed in a given population and the level of human exposure to the substance. This tool does not quantify the risk of a particular substance, but it indicates the level of safety concern of a contaminant present in food.[21] For glycidol, it was considered that a MOE of 25 000 or lower represents a health concern. The most alarming scenario was found for infants receiving infant formula only. The MOE found for this age group was 5500 for the mean occurrence level and 2100 for the 95th percentile of occurrence.[1]

5.3 Formation Conditions and Mechanisms

3-MCPDE, 2-MCPDE and GE are not present in quantifiable amounts in crude oils. These substances are formed during the refining process, particularly, under the deodorization conditions which apply high temperatures, usually above 200 °C.[22]

Most edible oils require a refining process in order to remove undesirable compounds that may affect the oil quality, such as phospholipids, FFA, pigments, volatiles, and contaminants. This process can be carried out either chemically or physically. The primary difference between them is that chemical refining has a neutralization step to remove FFA which includes an alkali treatment to neutralize the oil, whereas in physical refining FFA are eliminated by stripping during the deodorization step.[23]

The main steps of chemical refining are degumming, alkaline neutralization, followed by washing with water, bleaching and deodorization, while physical refining encompasses degumming, bleaching, and deodorization.[24,25] The choice of the type of refining will depend on the individual characteristics of the crude oil such as FFA and phospholipid content as well as the costs of the process.[25]

The deodorization is the most critical step of the oil refining, in which odoriferous compounds such as FFA, aldehydes, ketones, peroxides, alcohols and other organic compounds are removed from the oil. Deodorization consists of a stripping process in which a given amount of a stripping agent (usually steam) is injected through the oil at high temperatures and low pressures.[23,26] Deodorization has four main variables that can be adjusted by the refineries according to the type of oil and the desired final product, and these variables include temperature, pressure, stripping steam and time.[27]

Due to the high temperatures applied in the deodorization step (up to 260 °C), some undesirable changes can occur in the oil affecting negatively the quality of the final product such as the formation of *trans* fatty acids and an increase in interesterification reaction. In addition, 3-MCPDE, 2-MCPDE and GE are mainly formed in this step given the application of high temperatures.[3,22,28,29] The temperature is a factor that plays an important role in the formation of these compounds. Quantifiable levels of 3-MCPDE and GE in palm oil can be observed from 140 °C.[4] In addition, above 250 °C the concentration of GE increases exponentially with time.[22]

Besides the temperature, which is considered a critical variable in the formation of these contaminants, the influence of other factors has also been reported, such as the presence of chlorinated compounds in the oil that act as precursors of 3-MCPDE and 2-MCPDE. In vegetable oils, these compounds can be originated mainly from the endogenous metabolism of the plant and from the use of fertilizers containing chloride salts in cultivation. Moreover, the use of HCl-activated clays during the bleaching step of the refining process can also represent a source of chlorine.[30] At

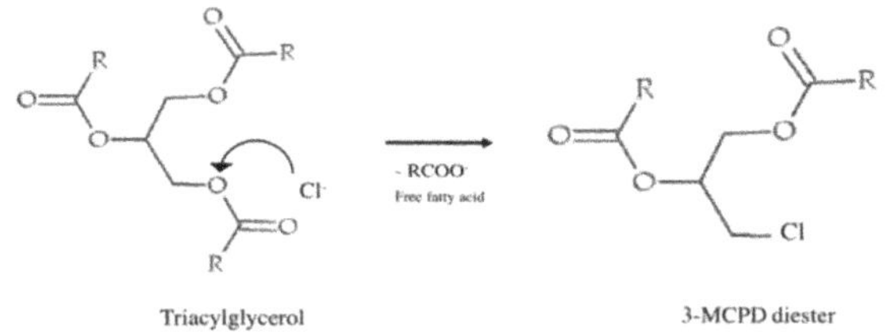

Figure 5.2 Nucleophilic attack of chloride in the triacylglycerol structure. Adapted from ref. 31 with permission from John Wiley and Sons, Copyright © 2011 WILEY-VCH Verlag GmbH & Co. KGaA, Weinheim.

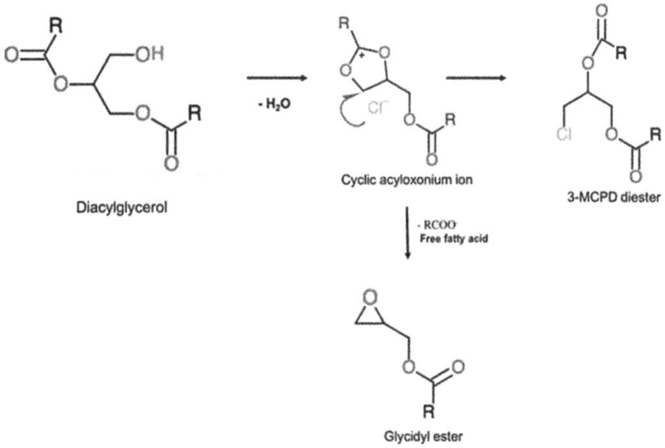

Figure 5.3 Formation of 3-MCPDE and GE by an intermediate cyclic acyloxonium ion. Adapted from ref. 31 with permission from John Wiley and Sons, Copyright © 2011 WILEY-VCH Verlag GmbH & Co. KGaA, Weinheim.

high temperatures, the chloride present in the oil, through nucleophilic substitution, attacks the glycerol carbon resulting in fatty acid elimination and formation of 3-MCPDE (Figure 5.2). Alternatively, the formation of 3-MCPDE may result from a nucleophilic attack of chlorides in intermediate ions as a cyclic acyloxonium (Figure 5.3).[31]

The presence of partial acylglycerols such as diacylglycerols (DAG) and monoacylglycerols (MAG) has been associated with the formation of GE.[32] The proposed mechanism involves intramolecular rearrangement that results in fatty acid elimination (Figure 5.4).[33] In addition, cyclic acyloxonium ions can be formed at deodorization conditions and initiate the formation of GEs (Figure 5.3).[34] Furthermore, it has been suggested that an interconversion between 2-MCPDE and 3-MCPDE through a GE intermediate can occur at high temperatures (Figure 5.5).[31] Also, it has been proposed that 3-MCPDE can be generated through a reaction mediated by free radicals (Figure 5.6).[35]

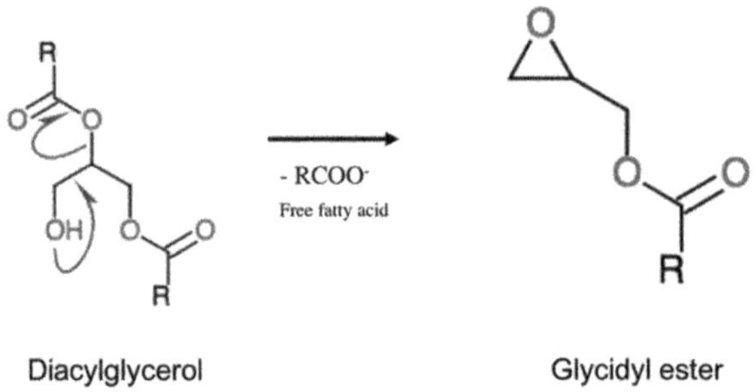

Figure 5.4 GE formation by an intramolecular rearrangement. Reproduced from ref. 33 with permission from Taylor & Francis (http://www.tandfonline.com), Copyright 2012.

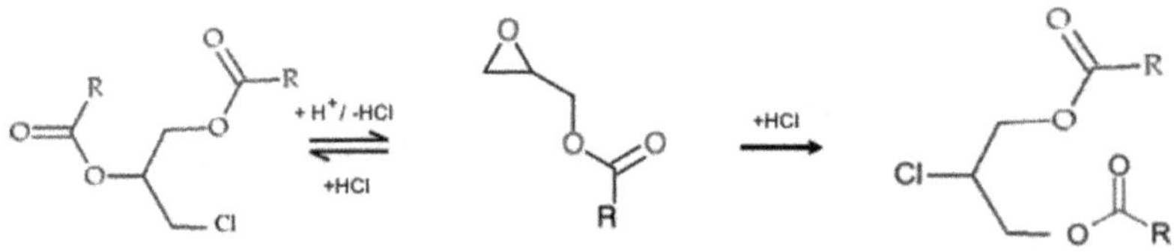

Figure 5.5 Interconversion between 3-MCPDE and 2-MCPDE. Adapted from ref. 31 with permission from John Wiley and Sons, Copyright © 2011 WILEY-VCH Verlag GmbH & Co. KGaA, Weinheim.

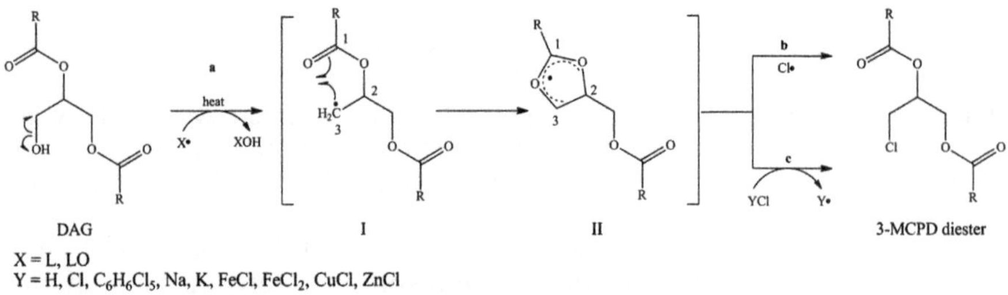

Figure 5.6 Proposed mechanism for the formation of 3-MCPD diesters from DAG mediated by a free radical. Reproduced from ref. 35 with permission from American Chemical Society, Copyright 2013.

5.4 Occurrence in Oils, Fats and Foods

The highest amount of 3-MCPDE, 2-MCPDE and GE is found in edible oils, particularly, in refined oils. Palm and palm-derived oils (palm olein and palm stearin) present the highest concentrations of these contaminants (Figure 5.7) due to the presence of precursors in the crude oil, such as chlorinated compounds and DAGs.[4,36]

Palm oil is an edible oil extracted from the mesocarp of the fruit of the *Elaeis guineensis* Jacq., which has a semisolid appearance at room temperature and reddish color due to its high content of carotenoids. This oil has a

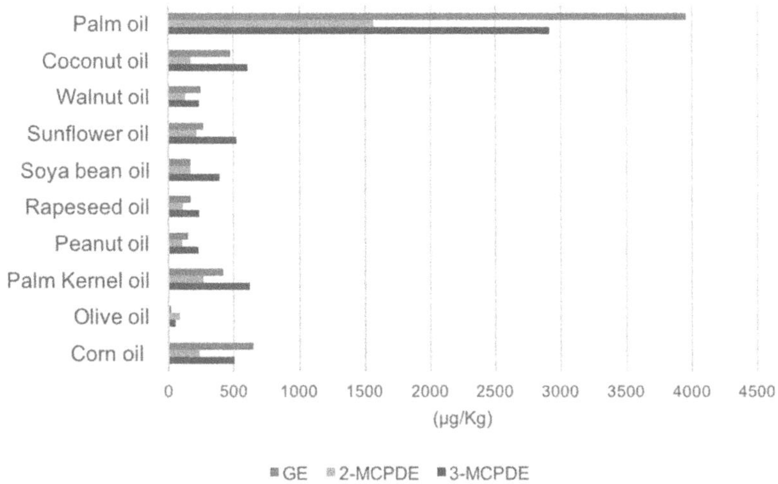

Figure 5.7 Concentration of 3-MCPDE, 2-MCPDE, and GE in different types of edible oils.

ratio of approximately 50% of saturated fatty acids and 50% of unsaturated fatty acids. The main fatty acids present in this oil are palmitic (43.4%), oleic (40%) and linoleic (10.5%).[37] Palm oil can be naturally fractionated in a liquid fraction (65–70%), known as palm olein, and in a solid fraction (30–35%), known as palm stearin. The typical composition of palm oil is 87–92% triacylglycerols (TAG), 3–8% DAG, 0–0.5% MAG and 1–5% FFA.[38,39]

Both refined palm oil (RPO) and its fractions have been widely applied in the food industry due to their functional benefits, versatility and low cost. These benefits include stability at high temperatures and during storage, excellent sensory attributes such as neutral odor and flavor, and creamy texture. RPO applications include the production of margarine, ice creams, wafers, fried foods, and infant formulas, among others.[40]

MCPD and glycidyl esters are not expected to be found in unrefined oil.[29] Extra virgin olive oil (EVOO), for example, is a cold-pressed oil extracted from fresh olives by mechanical means. This process does not require high temperatures and, consequently, MCPD and glycidyl esters are not expected in this product. However, it has been observed that these contaminants are present in EVOO commercialized in the market. A study conducted in Brazil showed that the levels of 3-MCPDE, 2-MCPDE and GE in EVOO were up to 1.16, 0.58 and 1.98 mg kg^{-1}, respectively.[41] Another work conducted in Taiwan reported levels of 3-MCPDE in EVOO up to 0.24 mg kg^{-1}.[42] Adulteration of EVOO accounts for about 24% of all food fraud records.[43] In this way, MCPDE and GE content could be used as a complementary indicator of adulteration in EVOO.

Table 5.1 shows that these contaminants also occur in various processed foods including meat products (mostly delicatessen products), dairy

Table 5.1 Occurrence of 3-MCPDE, 2-MCPDE and GE in different foodstuffs and oils. Reproduced from ref. 56 with permission from Elsevier, Copyright 2018.[a]

Food group		Number of samples	3-MCPDE ($\mu g\ kg^{-1}$)		2-MCPDE ($\mu g\ kg^{-1}$)		GE ($\mu g\ kg^{-1}$)		Method	Reference
			Mean	Range	Mean	Range	Mean	Range		
Carbohydrate-rich products	Potato chips	5	431.4	< LOQ–604	NR	NR	1.9	ND–9.5	GC-MS	49
	Corn puffs	4	195	45–267	NR	NR	ND	—		
	Sticks	5	318.5	25–257	NR	NR	2.3	< LOQ–11.6		
	Crackers	5	449.4	112–748	NR	NR	ND	—		
	Peanuts	3	475.3	251–753	NR	NR	ND	—		
	Granola	3	375.0	206–513	NR	NR	9.6	ND–28.8		
	Muesli	3	353.6	86–585	NR	NR	ND	—		
	Flakes	3	50.3	26–78	NR	NR	ND	—		
	Sugar-free biscuits	5	599.2	133–1501	NR	NR	ND	—		
	Organic farming biscuits	5	237.0	59–495	NR	NR	ND	—		
	Gluten-free biscuits	4	326.3	91–571	NR	NR	ND	—		
	Baby biscuits	6	283.5	88–443	NR	NR	ND	—		
	Classic biscuits	5	590	363–870	NR	NR	ND	—		
	Bread and bread rolls	75	29	23–36	14	9.8–19	8	7.8–8.3		
	Breakfast cereals	66	26	19–33	15	10–20	17	16–18	GC-MS	1
	Fine bakery wares	88	172	167–178	87	82–92	112	112–113		

Edible oils and fats	Olive oil	15	4.0	1.0–7.6	—	—	—	—	GC-MS/MS	42
	Olive pomace oil	7	12.3	7.4–20.53	—	—	—	—		
	Sunflower oil	11	230	80–960	110	20–520	230	20–900	GC-MS/MS	50
	Refined palm oil	6	1330	180–2480	650	80–1650	1870	100–3550		
	Refined rape seed oil	5	440	30–510	210	10–310	310	10–1100		
	Fish oil	5	—	1500–5500	—	100–230	—	—	GC-MS	51
	Margarine	5	—	1300–7300	—	630–1700	—	—		
	Corn oil	38	503	502–505	233	—	650	647–654	GC-MS	1
	Olive oil	9	48	48–49	86	85–88	15	0–31		
	Palm kernel oil	97	624	—	270	249–291	421	415–428		
	Peanut oil	8	229	—	102	90–115	148	133–162		
	Rapeseed oil	294	232	224–239	109	78–140	166	144–188		
	Soya bean oil	191	394	392–396	167	159–175	171	157–186		
	Sunflower oil	596	521	517–524	218	207–229	269	259–279		
	Coconut oil/fat	204	608	608	169	143–194	476	472–479		
	Palm oil/fat	501	2912	2912	1565	1563–1566	3955	3954–3955		
	Margarine and similar products	170	408	406–409	159	152–166	361	358–364		
	Extra virgin olive oil	46	133	ND–116	61	ND–580	323	ND–198	GC-MS	41
	Olive oil	13	855	280–3777	420	170–1910	643	ND–1880		
	Oil blends	17	304	180–610	120	ND–250	825	310–1840		
Infant formula	Infant formula	40	150	ND–630	—	—	220	ND–750	GC-MS	46
	Infant formula	98	370	24–920	—	—	84.4	< LOQ–400	LC-MS/MS	52
	Infant formula milk powder	88	185	0–136	41	0–52	—	—	GC-MS	53
Beef flavoring products		6	256.3	30.6–501.7	NR	NR	NR	NR	UPLC-TQMS	54
Pork, beef, chicken	Cooked by gas fired frying pan	—	—	—	—	—	70–170	NR	LC-MS/MS	55
	Cooked by charcoal grill	—	—	—	—	—	67–1110	NR		

[a]NR, Not reported in the study; ND, not detected; LOQ, limit of quantification; GC, gas chromatography; LC, liquid chromatography; MS, mass spectrometry; UPLC–TQMS, ultra-performance liquid chromatography-triple quadrupole mass spectrometry.

products, cereal, and bakery products, fries, soups, sauces, roasted coffee, and infant formulas.[1,44,45] The presence of significant amounts of MCPDE and GE in infant formulas has been raising potential concern.[1,46,47]

It has been reported that these esters are not formed endogenously in foods during processing, such as in fried foods, suggesting that these products are mainly contaminated by the oil used in frying, through a clear carry-over of the contaminants from the oil to the food.[48]

5.5 Mitigation Strategies

In order to reduce the formation of these contaminants, some strategies have been proposed with different approaches (Figure 5.8) that include: the removal of precursors, such as chlorinated compounds present in the crude oil; modifications of the processing parameters, in order to modify the drastic conditions applied in the refining process; the addition of refining aids to prevent the formation of the contaminants during processing; and degradation or removal of the contaminants formed in the final product using adsorbent agents.[8,9] Also, strategies to prevent the carry-over of the contaminants in fried foods due to oil uptake have been investigated. These measures will be further discussed in detail hereinafter.

It should be mentioned that the Codex Committee on Contaminants in Foods (CCCF) has been working on the elaboration of a code of practice for

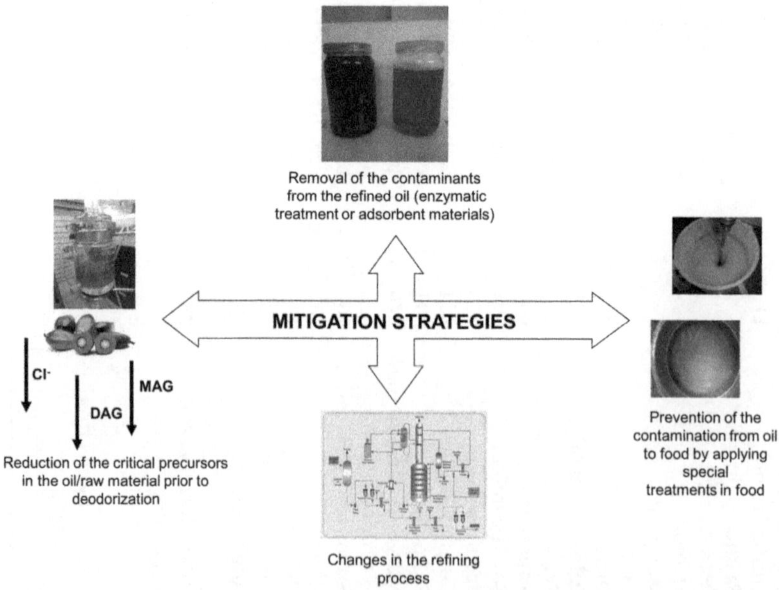

Figure 5.8 The main strategies reported to reduce 3-MCPDE, 2-MCPDE and GE levels in edible oils and foods.

the reduction of 3-MCPDE and GE in refined oils and food products made with refined oils. The document covers a series of mitigation measures in different productions stages of edible oils that includes harvesting, milling and refining.[57]

5.5.1 Removal of Precursors

Satisfactory results have been obtained by including a washing step of the oil prior to deodorization at the first stages of oil processing using an aqueous solution containing ethanol. This process is associated with the removal of organic chlorinated compounds and showed reductions of up to 38% in the final levels of 3-MCPD esters. Washing the palm fruit pulp prior to oil extraction yielded even more satisfactory results, around a 95% reduction in 3-MCPDE.[8,58] Similar results have been achieved with other types of oils. For instance, washing peanut oil with ethanol solution 50% (v/v) before deodorization reduced 3-MCPDE levels by 42.3%. As suggested in some studies, washing the oil at the early stages of palm oil refining could be more effective to remove the chlorinated precursors since these compounds at that stage are in smaller amounts and have lower lipophilicity.[59]

Washing bleached oil before deodorization has also been investigated and could be a potential alternative strategy to remove chlorinated compounds when HCl-activated bleaching earths are used during refining. Washing bleached peanut oil with ethanol solution using different ratios of volumes before deodorization showed a reduction in the DAG levels and, consequently, a reduction in GE levels.[32] In experiments performed by our research group (unpublished results), a sodium thiosulfate aqueous solution was tested to wash bleached palm oil (BPO) before deodorization in order to mitigate MCPDE and GE. The use of this salt could be a strategy to remove chlorinated compounds in the oil given its high reactivity with chlorine and the fact that the thiosulfate ion behaves as a good nucleophile that could attack alkyl halides, releasing halides in the reaction medium. If such an attack occurred on the chlorinated precursors present in the BPO in the form of alkyl halide, it could result in a reduction in the level of chlorinated precursors since the chlorine resulting from that reaction would be soluble in the aqueous phase of the partition. However, it was observed that washing BPO with an aqueous solution of sodium thiosulfate (2.5%) and an ethanolic solution of sodium thiosulfate (2.5%) led to a slight increase in 3-MCPDE, 2-MCPDE and GE levels compared with control. The possible explanation for this outcome is a salting-out effect caused by sodium thiosulfate that saturated the washing solution preventing it from solubilizing the water- and ethanol-soluble chlorinated compounds. Also, the addition of the salt seems to favor the formation of DAGs, resulting in an increase in the contaminants (Figure 5.9).[60]

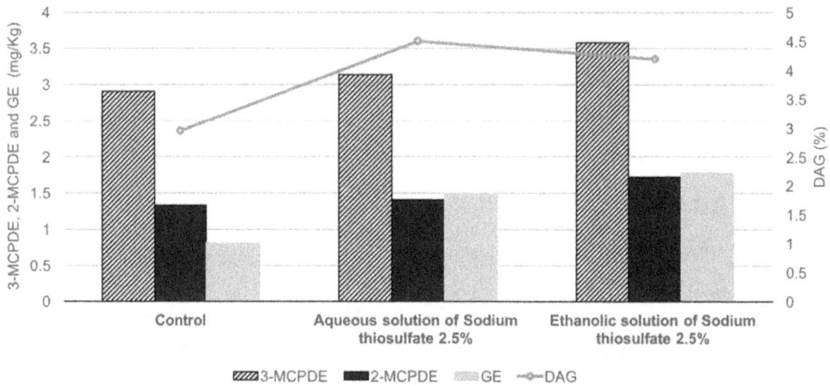

Figure 5.9 Effect of washing bleached palm oil (BPO) with sodium thiosulfate on the formation of 3-MCPDE, 2-MCPDE, and GE.

5.5.2 Modification of Processing Parameters

Reduction in the levels of the contaminants during refining can be achieved by reducing the harsh conditions used in the deodorization step, illustrated by the high temperature and the deodorization time which have a decisive influence on the formation of 3-MCPDE, 2-MCPDE and GE.[26]

Different alternatives have been suggested to moderate the deodorization conditions and one of them is the method known as dual deodorization. This method combines a short time deodorization step at a higher temperature with a long deodorization step at a low temperature. Usually, the first step is carried out at 250–270 °C for a short time and a prolonged time at 200 °C.[61] This approach has been applied by industries to preserve tocopherol (fat-soluble vitamin E) and tocotrienol in the oil and reduce *trans* fatty acid formation.[62] Dual deodorization reduced the formation of 3-MCPDE and GE by up to 80% compared with conventional deodorization.[61]

Another possible way of reducing the formation of the contaminants by modifying the conditions of the deodorization step is the use of short-path distillation. This is a distillation technique that involves the distillate moving through a short distance at a reduced pressure of 0.001–1 mbar, which allows lowering of the boiling point of substances. This technique is applied when heat-sensitive components have to be distilled or to prevent the oil from physical and chemical changes that occur at high temperatures such as the formation of *trans* fatty acids.[63]

It has been observed that short-path distillation produced a refined palm oil with no 3-MCPDE and low GE. However, there was a substantial loss in sensory attributes in the refined oil. Thus, this result limits the application of this technique for commercial purposes.[64] Furthermore, short-path distillation has some limitations in terms of up-scaling the refining process and it can increase the costs of the process.

5.5.3 Addition of Refining Aids

Another approach used to mitigate the formation of the contaminants is the addition of substances that can prevent their formation or compete with precursors for reaction with the chlorine-donating compound during deodorization.

Adding glycerol and ethanol in the oil as a refining aid could achieve ~25–35% reduction in 3-MCPDE levels. This hypothesis is based on the ability of ethanol to form volatile chlorinated adducts that would be removed from the oil during deodorization.[58]

The addition of diacetin, a short-chained DAG, has been investigated since it could react with chlorinated compounds and be subsequently removed from the process due to its low boiling point.[8] It has been demonstrated that diacetin could reduce the formation of 3-MCPDE during peanut oil refining by 43.8, 46.4 and 44.6% when using a weight ratio of diacetin of 4, 8, and 10%, respectively.[32]

Another possibility is the addition of a substance that can neutralize FFAs before deodorization such as carbonates and bicarbonates. Experimental data from Ŝmidrkal *et al.* (2016) showed that alkaline carbonates or hydrogen carbonates added before deodorization can completely avoid the formation of 3-MCPDE.[65]

Also, the addition of formic acid in the water used to generate the stripping steam for the deodorization presented a reduction in GE levels by 20 and 40% when using a concentration of 5 and 10% of formic acid, respectively.[66]

5.5.4 Removal of the Contaminants from the Refined Oil

Another mitigation approach is the removal of the contaminants after the refining process, by means of adding an agent that can selectively react with the contaminants resulting in non-toxic products.

Some studies have investigated the ability of different inorganic adsorbent materials to reduce 3-MCPDE levels in refined palm oil. The following materials have been tested: amorphous magnesium silicate, magnesium silicate, zeolite, silicon oxide, sodium aluminum silicate, synthetic calcium silicate, and synthetic magnesium silicate. Calcinated zeolite and a synthetic magnesium silicate were the only materials that could reduce the content of 3-MCPDE, with a reduction of up to 40%.[67]

The removal of 3-MCPDE from refined oils by an enzymatic treatment has been evaluated as well. The first step of the experiments consisted of the hydrolysis of the esters by *Candida antarctica* lipase A, releasing free 3-MCPD. Then, the degradation of free 3-MCPD was achieved by the combination of the enzyme halohydrin dehalogenase from *Arthrobacter* sp. and an epoxide hydrolase from *Agrobacterium radiobacter* converting free 3-MCPD into the non-toxic product glycerol *via* glycidol. This enzymatic treatment could reduce the levels of 3-MCPD by up to 70%.[68]

5.5.5 Reducing Contamination from the Oil to the Food

Strategies to minimize the contamination of fried foods have also been investigated. These strategies focus on using material or treatments in the food during or prior to frying to prevent contaminated oil uptake in food.

The effect of pre-treatments on the contamination of French fries by 3-MCPDE has been investigated. It was observed that a pectin coating in blanched potatoes reduced the levels of 3-MCPDE by 19%. The reduction is possibly due to the formation of a calcium–pectate complex on the potato resulting in less oil absorption and an increase in stiffness of the cell wall of the potato, which consequently reduced the contamination by 3-MCPDE.[69]

Strategies on the reduction of contamination of pre-fried breaded and frozen fish products have been studied. These strategies included the addition of an adsorbent material (zeolite) to the pre-frying oil, as well as the addition of L-cysteine in the breading ingredient. Both strategies could reduce 3-MCPDE, 2-MCPDE and GE levels significantly in fried fish.[70]

5.6 Analytical Methods

The analytical approaches for the determination of 3-MCPDE, 2-MCPDE and GE in foodstuffs involve: indirect analysis, in which the total concentration of the contaminants is measured as free 3-MCPD, 2-MCPD and glycidol obtained after a hydrolysis/methanolysis of the esters; and direct analysis, in which the 3-MCPD, 2-MCPD and glycidyl esters are individually identified.[71]

The first step of indirect analysis is the cleavage of esters to their free forms (3-MCPD, 2-MCPD and glycidol) in the oil sample or the lipid phase of foods. This can be carried out through a transesterification reaction in the presence of methanol either in acid or alkaline media.[72] The transesterification reaction is followed by neutralization and salting-out steps and then the analytes are derivatized with phenylboronic acid to obtain highly volatile derivatives before injection into a gas chromatography (GC) system.

Three official methods have been proposed by the American Oil Chemists' Society (AOCS): Cd 29a-13, Cd 29b-13, and Cd 29c-13. These methods are based on the indirect determination of these contaminants in oils and consist of a simultaneous determination of 3-MCPDE, 2-MCPDE, and GE. In the method Cd 29a-13, GE is converted into 3-monobromopropanediol fatty acid esters (3-MBPDE) in an acid solution of sodium bromide. Also, 3-MBPDE (formed from GE), 3-MCPDE and 2-MCPDE are then converted into the non-esterified forms in an acid methanolic solution. The methods Cd 29b-13 and Cd 29c-13 employ a base catalysis using an alkaline alcoholic solution to release the free forms of the analytes. The method Cd 29c-13 involves differential measurement of GE by the application of two protocols and does not quantify 2-MCPDE. In 2017, the AOCS published a new protocol, Cd 30-15, which is used for the analysis of contaminants in emulsions.[73,74]

The primary advantages of indirect methods are the low number of analytical standards and a more straightforward interpretation of results. The disadvantages include the several steps that must be carried out before injection into a GC system such as transesterification, neutralization, salting-out and derivatization.[56]

The direct analysis determines the esters without chemical reactions such as transesterification and derivatization, reducing the time of analysis. Thus, they are determined as they are found in foods. Usually, high performance liquid chromatography coupled to mass spectrometry (HPLC-MS) is the selected technique for the determination of the individual compounds, using time of flight (TOF), Orbitrap™ and triple quadrupole mass spectrometer (MS/MS).[2] The disadvantage of the method is the large number of analytical standards that have to be included in the analysis resulting in increased costs of the assay.[71]

5.7 Regulatory Aspects

Based on risk assessment results of 3-MCPDE and GE, Regulation (EU) 2018/290 amending Regulation (EU) 1881/2006 was recently adopted across the European Union as a risk management strategy. In this regulation, the European Commission published the maximum levels for GE in vegetable oils and fats, infant formula, follow-on formula and foods intended for medical and baby nutrition. For vegetable oils and fats placed on the market for the final consumer or for use as an ingredient in food, with the exception of foods intended for baby nutrition, the limit set was of 1000 $\mu g\ kg^{-1}$. For vegetable oils and fats destined for the production of baby food and processed cereal-based food for infants and young children, the limit set was of 500 $\mu g\ kg^{-1}$. For infant formula, follow-on formula and foods for special medical purposes intended for infants and young children, the limits set until June of 2019 were of 75 $\mu g\ kg^{-1}$ (powdered food) and 10 $\mu g\ kg^{-1}$ (liquid food). After this date, new limits will be of 50 $\mu g\ kg^{-1}$ (powdered food) and 6 $\mu g\ kg^{-1}$ (liquid food).[75]

No limits have yet been set for 3-MCPDE and 2-MCPDE in refined oils and foodstuffs, but a regulation for these contaminants should eventually be proposed in the future.[19,75]

5.8 Conclusion

The edible oil and food industries have been facing a significant challenge in achieving an effective reduction in the levels of 3-MCPDE, 2-MCPDE and GE in order to provide a safe product to consumers. Different mitigation approaches have been investigated so far, but the great challenge is to upscale these strategies without a negative impact on the oil quality. It is quite clear that the way forward for this issue is the implementation of various techniques at different stages of the oil processing chain. This approach could be similarly compared with the hurdle technology applied in food microbiology which consists of combining methods or processes at various points in the food production chain, which secures the safety, stability, organoleptic properties, nutritional quality and the economic viability of food products. Regulations recently adopted reinforce the idea that it is of the utmost importance to implement mitigation strategies to reduce 3-MCPDE, 2-MCPDE and GE in foods and food ingredients.

Acknowledgements

The authors acknowledge the São Paulo Research Foundation (FAPESP) [grant numbers 2016/239583, 2017/26675-5 and 2014/2152-0]. W.C.S. thanks the financial support of Brazilian Federal Agency for Support and Evaluation of Graduate Education within the Ministry of Education of Brazil (CAPES) under Grant number 33003017027P1 – PROEX0082040.

References

1. European Food Safety Authority (EFSA), *EFSA J.*, 2016, **14**, 4426.
2. A. P. Arisseto, P. F. C. Marcolino, E. Vicente and K. A. Sampaio, *Quim. Nova*, 2013, **36**, 1406–1415.
3. Z. Zelinková, B. Svejkovská, J. Velísek and M. Dolezal, *Food Addit. Contam.*, 2006, **23**, 1290–1298.
4. W. W. Cheng, G. Q. Liu, L. Q. Wang and Z. S. Liu, *Compr. Rev. Food Sci. Food Saf.*, 2017, **16**, 263–281.
5. A. J. Dijkstra and J. C. Segers, in *The Lipid Handbook*, ed. F. D. Gunstone, J. L. Harwood and A. J. Dijkstra, CRC Press, Boca Raton, 3rd edn, 2007, ch. 3, p. 120.
6. F. Pudel, P. Fehling, K. Vosmann, A. Freudenstein and B. Mattha, *Eur. J. Lipid Sci. Technol.*, 2011, **113**, 380–386.
7. K. A. Sampaio, A. P. Arisseto, J. V. Ayala, C. Stevens, R. Ceriani, R. Verhé and A. J. Meirelles, *Toxicol. Lett.*, 2013, **221S**, 9–18.
8. B. Matthäus and F. Pudel, *Lipid Technol.*, 2013, **25**, 151–155.
9. M. R. Ramli, W. L. Siew, N. A. Ibrahim, A. Kuntom and R. A. Abd. Razak, *Food Addit. Contam., Part A*, 2015, **32**, 817–824.
10. International Agency for Research on Cancer (IARC), 3-Monochloro-1,2-propanediol, https://monographs.iarc.fr/wpcontent/uploads/2018/06/mono101-010.pdf, accessed November 2018.
11. J. Orsavova, L. Misurcova, J. Vavra Ambrozova, R. Vicha and J. Mlcek, *Int. J. Mol. Sci.*, 2015, **16**, 12871–12890.
12. K. Abraham, K. E. Appel, E. Berger-Preiss, E. Apel, S. Gerling, H. Mielke, O. Creutzenberg and A. Lampen, *Arch. Toxicol.*, 2013, **87**, 649–659.
13. International Agency for Research on Cancer (IARC), Glycidol, https://monographs.iarc.fr/wp-content/uploads/2018/06/mono77-19.pdf, accessed November 2018.

14. F. Frenzel, A. Oberemm, A. Braeuning and A. Lampen, *Food Chem. Toxicol.*, 2018, **121**, 1–10.
15. G. Huang, B. Gao, J. Xue, Z. Cheng, X. Sun, Y. Zhang and L. L. Yu, *J. Agric. Food Chem.*, 2018, **66**, 11672–11680.
16. J. Aasa, M. Törnqvist and L. Abramsson-Zetterberg, *Food Chem. Toxicol.*, 2017, **109**, 414–420.
17. G. Huang, J. Xue, X. Sun, J. Wang, Y. Wang, T. T. Y. Wang and L. Y. Liangli, *Toxicol. in Vitro*, 2018, **51**, 54–62.
18. A. Segal, J. J. Solomon and F. Mukai, *Cancer Biochem. Biophys.*, 1990, **11**, 59–67.
19. European Food Safety Authority (EFSA), *EFSA J.*, 2018, **16**, 1–48.
20. World Health Organization (FAO/WHO), *WHO Techincal Report Series*, 2017, **1002**, 74–105.
21. European Food Safety Authority (EFSA), *EFSA J.*, 2012, **10**, 2578.
22. F. Pudel, P. Benecke, P. Fehling, A. Freudenstein, B. Matthäus and A. Schwaf, *Eur. J. Lipid Sci. Technol.*, 2011, **113**, 368–373.
23. R. D. O. Brien, *Oils and Fats: Formulating and Processing for Applications*, CRC Press, Boca Raton, 2008.
24. M. Kellens and W. De Greyt, in *Introduction to Fats and Oil Technology*, ed. P. J. O'Brien, R. D. Farr and W. E. Wan, AOCS Press, Champaign, 2nd edn, 2000, ch. 13, pp. 235–268.
25. C. Vaisali, S. Charanyaa, P. D. Belur and I. Regupathi, *Int. J. Food Sci. Technol.*, 2014, **50**, 13–23.
26. K. A. Sampaio, J. V. Ayala, V. Van Hoed, S. Monteiro, R. Ceriani, R. Verhé and A. J. A. Meirelles, *J. Food Sci.*, 2017, **82**, 1842–1850.
27. R. Ceriani and A. J. A. Meirelles, *J. Am. Oil Chem. Soc.*, 2004, **81**, 305–312.
28. K. Franke, U. Strijowski, G. Fleck and F. Pudel, *LWT--Food Sci. Technol.*, 2009, **42**, 1751–1754.
29. K. Hrncirik and G. van Duijn, *Eur. J. Lipid Sci. Technol.*, 2011, **113**, 374–379.
30. B. D. Craft and F. Destaillats, in *Processing Contaminants in Edible Oils*, ed. S. MacMahon, Elsevier, Urbana, 1st edn, 2014, ch. 1, pp. 7–21.
31. A. K. K. Rahn and V. A. Yaylayan, *Eur. J. Lipid Sci. Technol.*, 2011, **113**, 323–329.
32. C. Li, L. Li, H. Jia, Y. Wang, M. Shen, S. Nie and M. Xie, *Food Chem.*, 2016, **199**, 605–611.
33. F. Destaillats, B. D. Craft, L. Sandoz and K. Nagy, *Food Addit. Contam., Part A: Chem., Anal., Control, Exposure Risk Assess.*, 2012, **29**, 29–37.
34. R. Weißhaar and R. Perz, *Eur. J. Lipid Sci. Technol.*, 2010, **112**, 158–165.
35. X. Zhang, B. Gao, F. Qin, H. Shi, Y. Jiang and X. Xu, *J. Agric. Food Chem.*, 2013, **61**, 2548–2555.
36. S. H. Tiong, N. Saparin, H. F. Teh, T. L. M. Ng, M. Z. Bin Md Zain, B. K. Neoh, A. Md Noor, C. P. Tan, O. M. Lai and D. R. Appleton, *J. Agric. Food Chem.*, 2018, **66**, 999–1007.
37. D. O. Edem, *Plant Foods Hum. Nutr.*, 2002, **57**, 319–341.

38. A. Szulczewska-Remi, M. Nogala-Kalucka, J. Kwiatkowski, E. Lampart-Szczapa and M. Rudzinska, *J. Food Lipids*, 2005, **12**, 112–123.
39. K. A. Sampaio, PhD thesis, Universidade Estadual de Campinas, 2011. http://repositorio.unicamp.br/jspui/bitstream/REPOSIP/255200/1/Sampaio_KliciaAraujo_D.pdf, accessed November 2018.
40. M. E. Norhaizan, S. Hosseini, S. Gangadaran, S. T. Lee, F. R. Kapourchali and M. H. Moghadasian, *Lipid Technol.*, 2013, **25**, 39–42.
41. K. Kamikata, E. Vicente, A. P. Arisseto-Bragotto, A. M. R. O. Miguel, R. F. Milani and S. A. V. Tfouni, *Food Control*, 2019, **95**, 135–141.
42. W. C. Hung, G. J. Peng, W. J. Tsai, M. H. Chang, C. D. Liao, S. H. Tseng, Y. M. Kao, D. Y. Wang and H. F. Cheng, *Food Addit. Contam., Part B*, 2017, **10**, 233–239.
43. Food Fraud and "Economically Motivated Adulteration" of Food and Food Ingredients, https://fas.org/sgp/crs/misc/R43358.pd, accessed November 2018.
44. C. G. Hamlet, P. A. Sadd, C. Crews, J. Velíšek and D. E. Baxter, *Food Addit. Contam.*, 2002, **19**, 619–631.
45. B. Svejkovska, O. Novotny, V. Divinova, Z. Reblova, M. Dolezal and J. Velisek, *Czech J. Food Sci.*, 2004, **22**, 190–196.
46. A. P. Arisseto, W. C. Silva, G. R. Scaranelo and E. Vicente, *Food Control*, 2017, **77**, 76–81.
47. J. H. Spungen, S. MacMahon, J. Leigh, B. Flannery, G. Kim, S. Chirtel and D. Smegal, *Food Addit. Contam., Part A: Chem., Anal., Control, Exposure Risk Assess.*, 2018, **35**, 1085–1092.
48. A. P. Arisseto, P. F. C. Marcolino, A. C. Augusti, G. R. Scaranelo, S. A. G. Berbari, A. M. R. O. Miguel, M. A. Morgano and E. Vicente, *J. Am. Oil Chem. Soc.*, 2017, **94**, 449–455.
49. A. Sadowska-Rociek, M. Surma and E. Cieslik, *Rocz. Panstw. Zakl. Hig.*, 2018, **69**, 127–137.
50. J. Kuhlmann, *Eur. J. Lipid Sci. Technol.*, 2016, **118**, 382–395.
51. R. Jędrkiewicz, A. Głowacz, J. Gromadzka and J. Namieśnik, *Food Control*, 2016, **59**, 487–492.
52. J. Leigh and S. MacMahon, *Food Addit. Contam., Part A: Chem., Anal., Control, Exposure Risk Assess.*, 2017, **34**, 356–370.
53. P. Wang, L. Ying, Y. Hu, Z. Wang, T. Shen and X. Wu, *J. AOAC Int.*, 2016, **99**, 786–791.
54. Q. Chai, Z. Xiaoming, E. Karangwa, Q. Dai, S. Xia, J. Yu and Y. Gao, *RSC Adv.*, 2016, **6**, 113576–113582.
55. R. Inagaki and C. Hirai, *J. Food Process. Technol.*, 2016, **07**, 2–7.
56. A. P. Arisseto, W. C. Silva, R. G. Tivanello, K. A. Sampaio and E. Vicente, *Curr. Opin. Food Sci.*, 2018, **24**, 36–42.
57. Report of the 12th session of the Codex Committee on Contaminants in Foods, http://www.who.int/foodsafety/areas_work/food-standard/CAC41/en/, accessed November 2018.
58. B. D. Craft, K. Nagy, L. Sandoz and F. Destaillats, *Food Addit. Contam., Part A*, 2011, **29**, 354–361.

59. K. Nagy, L. Sandoz, B. D. Craft and F. Destaillats, *Food Addit. Contam., Part A*, 2011, **28**, 1492–1500.
60. W. C. Silva, unpublished work, 2017.
61. B. Matthäus and F. Pudel, in *Processing Contaminants in Edible Oils*, ed. S. MacMahon, AOCS Press, Urbana, 2014, ch. 2, pp. 23–55.
62. V. Gibon, W. De Greyt and M. Kellens, *Eur. J. Lipid Sci. Technol.*, 2007, **109**, 315–335.
63. T. E. Suliman, J. Jiang and Y. Liu, *Int. J. Eng. Sci. Technol.*, 2013, **5**, 654–662.
64. F. Pudel, P. Benecke, K. Vosmann and B. Matthäus, *Eur. J. Lipid Sci. Technol.*, 2016, **118**, 396–405.
65. J. Šmidrkal, M. Tesařová, I. Hrádková, M. Berčíková, A. Adamčíková and V. Filip, *Food Chem.*, 2016, **211**, 124–129.
66. B. Matthäus, F. Pudel, P. Fehling, K. Vosmann and A. Freudenstein, *Eur. J. Lipid Sci. Technol.*, 2011, **113**, 380–386.
67. U. Strijowski, V. Heinz and K. Franke, *Eur. J. Lipid Sci. Technol.*, 2011, **113**, 387–392.
68. U. T. Bornscheuer and M. Hesseler, *Eur. J. Lipid Sci. Technol.*, 2010, **112**, 153–154.
69. A. P. Arisseto, *Food Res. Int.*, DOI: 10.1016/j.foodres.2018.10.070, in press.
70. S. Merkle, U. Ostermeyer, S. Rohn, H. Karl and J. Fritsche, *Food Chem.*, 2018, **245**, 196–204.
71. M. Dubois, A. Tarres, T. Goldmann, A. M. Empl, A. Donaubauer and W. Seefelder, *J. Chromatogr. A*, 2012, **1236**, 189–201.
72. V. Divinová, B. Svejkovská, M. Doležal and J. Velíšek, *Czech J. Food Sci.*, 2004, **22**, 182–189.
73. American Oil Chemists' Society (AOCS), *The Official Methods and Recommended Practices of the AOCS*, AOCS Press, 6th edn, 2013.
74. American Oil Chemists' Society (AOCS), *The Official Methods and Recommended Practices of the AOCS*, AOCS Press, 7th edn, 2017.
75. Commission Regulation (EU) 2018/290, https://eur-lex.europa.eu/legal-content/EN/TXT/PDF/?uri=CELEX:32018R0290&qid=, accessed November, 2018.

CHAPTER 6

Polycyclic Aromatic Hydrocarbons in Processed Food: Scientific Challenges and Research Recommendations

LOCHAN SINGH AND TRIPTI AGARWAL*

National Institute of Food Technology Entrepreneurship and Management Kundli, Sonipat, Haryana 131028, India
*E-mail: tripti.niftem@gmail.com

6.1 Introduction

Polycyclic aromatic hydrocarbons (PAHs) are lipophilic chemical compounds, a group of persistent organic pollutants and carcinogens, having two or more benzene rings.[1,2] Thousands of these compounds are present ubiquitously in different environmental matrices such as air,[3] soil,[4] water,[5] food,[6] ice[7] and interstellar space.[8] The mechanism for their formation and occurrence in these matrices commonly revolves around incomplete combustion of organic material, temperature-dependent adsorption–absorption and phase partitioning based on heterogeneous chemical reactions. The precise mechanism however differs with the nature of the matrix, its chemical composition and available reaction factors.[3] Among thousands

Food Chemistry, Function and Analysis No.19
Mitigating Contamination from Food Processing
Edited by Catherine S. Birch and Graham A. Bonwick
© The Royal Society of Chemistry 2020
Published by the Royal Society of Chemistry, www.rsc.org

of compounds falling into this category,[9] very few of them have been characteristically studied and evaluated.[2] The United States Environmental Protection Agency (US EPA) in 1976 selected 16 priority PAHs (16PAHs) which have been widely accepted and targeted in different research studies conducted worldwide in diverse environmental matrices. These include the compounds naphthalene (Naph), acenaphthylene (Acy), acenaphthene (Acen), phenanthrene (Phen), fluorene (Flu), benzo(g,h,i)perylene [B(ghi)P], anthracene (Anth), pyrene (Pyr), fluoranthene (Fluo), benzo(*a*)anthracene [B(*a*)A], benzo(*b*)fluoranthene [B(*b*)F], chrysene (Chry), benzo(*k*)fluoranthene [B(*k*)F], dibenzo(*a*,*h*)anthracene [DB(*ah*)A], benzo(*a*)pyrene [B(*a*)P] and indeno(1,2,3-*cd*)pyrene [IP] which have been further classified based on their ring structure. These compounds hold immense importance to human health and safety due to their toxic and carcinogenic nature as described by the International Agency for Research on Cancer (IARC).

Studies on the different aspects of PAHs have been a concern of research for the past 60 years and special importance has been given with respect to their presence in food products, their analysis and related human health risk assessments. PAHs in food is a consequence of contamination taking place at pre-harvest stages through contaminated environment (soil, water and air) and through the different processing and handling operations performed at post-harvest stages.[6] Often it has been observed that the toxicity associated with PAHs in food increases many fold after processing and cooking and thus is a highlighted concern. Despite all the advancements made in the past years to deal with the occurrence of PAHs in processed food, the field is still accompanied by scientific challenges and gaps. This chapter is thus aimed at discussing the diverse challenges present in our understanding about PAHs, their occurrence in processed food, detection, analytical advancements and health risk assessments. The chapter then discusses some of the critical gaps in present cognizance about PAHs in processed food products, such as lack of detailed mechanisms for their formation and reduction while processing and cooking, presence of PAH derivatives in food products and their toxicity, absence of diagnostic ratios for differentiating sources *etc.* Lastly, the chapter suggests recommendations based on the prevalent scientific background which will describe the course for future research studies.

6.2 General Concerns on PAH Toxicity

The combined effect of the PAH mixtures in a particular matrix has been found to vary from the predicted response based on simple addition of components. Also, most of the individual PAHs have been found to be non-carcinogenic or weakly carcinogenic with variable responses to tumorigenesis upon different exposure pathways. For example, naphthalene shows tumor formation in tissue upon inhalation but is probably non-genotoxic, its dietary role is not fully evaluated and still at an uncertain stage. It is also postulated that such low molecular weight PAHs, other than B(*a*)P, may act as promoters of oral and other carcinogenicity. It must also be noted here that for the data available for human exposure to PAHs, the effects and assessment through biomarkers are specific to occupation and environmental conditions. The lack of direct evidence and quantitative data for oral exposure to these compounds is an obvious gap in the knowledge base. This makes it difficult to use available data for dietary risk assessments. The basis utilized by the Joint FAO/WHO Expert Committee on Food Additives (JECFA) for toxicity evaluation thus consider the data available for tumour-bearing mice treated with coal tar mixtures. Besides the nature of PAH causing uncertainty in risk evaluation, the animal considered for evaluating PAH toxicity (upon dietary exposure) also exhibits variation in its tumour profile and development capacity, thereby raising alarm about the level of relevance within the human system. Despite the numerous studies and research data available on PAH toxicity, the effect of these compounds on metabolic responses in humans has not yet been determined. Studies directly focusing on metabolic changes associated with PAH exposure in humans are important as these compounds do not directly become carcinogenic but give rise to other indirect acting-carcinogenic compounds or pro-carcinogens within the human body upon metabolic activation. As such, it is important to study their metabolism in humans based on metabolomic analysis. As observed, these gaps in studies highlight that no human population sub-group has been identified as susceptible, owing to lack of human-specific and cell-based studies.

Previous toxicity-based studies have shown that PAH carcinogenesis begins with the formation of deoxy-ribose nucleic acid (DNA)-adducts, but the early stage detection of these adducts in whole tissue does not give a clear estimate of cancer risk. However, use of ^{2}P-post labeling does not distinguish clear results based on DNA-adducts within the cells with tumor growth and non-tumor growth, suggesting a role of additional critical factors. Accounting for these factors at early PAH carcinogenesis stages will prove beneficial in early cancer detection and prevention.[10] Hence, studying specific PAHs and adducts in cells may help in locating their point of action in cells/tissues and cannot be completely ignored. At present, DNA adduct quantification does not act as a good marker for PAH exposure as they may also be detected in non-targeted cells as stated above. Hence other biomarkers must be defined. Also, the tools to determine such changes are restricted to liquid chromatography mass spectroscopy–electron spray ionization (LC-MS-ESI) methods. Advanced techniques such as LC-MS-atmospheric pressure photo-ionization (APPI) needs further development. Sample analysis on gas chromatography (GC) equipment for PAH determination requires special attention to the type of injection mode, as the injection mode significantly affects the discrimination of compounds mainly of a less volatile nature. In this regard, an appropriate packed liner or multi-baffled liner proves beneficial. It must be noted that, most of the mass spectrometric applications use solid-phase extraction (SPE) processes where sorbents/fibers are used for purification. Thus, when using SPE processes for these compounds and adducts, care needs to be taken with the type of fibers involved (for example, polydimethylsiloxanes, glass wool *etc.*). Exploring other adsorbents and fiber materials may prove beneficial in this regard especially for heavy PAH compounds.[11]

Besides studying PAH-DNA adducts, telomere dysfunction and DNA methylation are some epigenetic changes reported for occupational PAH exposure in humans and in *in vivo* mouse studies.[12–14] These effects and changes warrant further research to gain further insights into PAH carcinogenesis and aid in understanding the precise mechanisms defining PAH carcinogenicity.[15] Lastly, continuous amendments to a priority PAH list, based on new insights available into oxygenated, nitro, alkylated and hydroxylated PAHs and newly identified parent PAHs, must be open for discussion at a global platform level and robust guidelines must be formulated since recent reports have also shown their presence in processed food.

6.3 Processed Food: A Challenge

Given the enormous diversity existing among food products, specific consumable species, ingredients used, cooking styles and methods prevalent throughout the world, huge variability is found within the single, predefined term 'processed food' which poses a challenge for food technologists and regulatory bodies while accounting for associated effects on the human body.[6] Understanding a single compound, parameter or phenomenon itself requires substantial effort and skill. Gaining a complete knowledge and obtaining a holistic view of food combining all parameters and influential factors thus becomes very difficult. Food items such as eggs, meat *etc.* which have been recommended as part of a healthy diet owing to their nutritional benefits, may contain PAHs and other toxicants before and after processing. Understanding the harmony between nutritional and harmful compounds present in food, while considering the beneficial effects to the human body of consuming such foods, and understanding the mechanism by which this balance may be maintained and altered to minimize levels of toxicants to safe limits, forms part of this challenge. Moreover, evaluating the changes in this balance after cooking and processing, due to variations occurring in the composition of raw food and processed food *i.e.* in terms of physiological and chemical parameters,[16] will be quite challenging. We must also note in the present scenario that the food product may simultaneously be affected by one or more pollutants or toxicants, which among themselves may result in synergistic or antagonistic effects[17,18] in the form of a cocktail or multi-residue toxic effect. In most studies to date, the direct effect of cooking on the physical and chemical characteristics of contaminants is not directly or relevantly correlated with adverse toxicological effects.[16] Thus, further studies are warranted to determine whether the observed variations, mainly in the reduction observed in PAH levels in these studies, occur as a result of the cooking type or are merely associated with the cooking conditions.

To date most of the studies in this area have been found to mainly focus on single cooking processes (one or more food products), single food product (one or more cooking processes), one or more variables and parent PAHs (number depending on guideline reference, detected compounds). The most extensively reported and studied data exist between food lipid content and PAH concentration, however the exact mechanism for this interaction is still not well characterized. The effects of protein and other components are given less importance and studies on these are limited in the published literature.[16] Only recently, Nie and coworkers[19] in their study on grilled pork sausages have reported that basic amino acids (L-arginine and L-lysine) and low molecular weight D-glucose favour PAH formation. They further suggest that while incorporating food additives during processing of food, it is necessary to systematically consider molecular properties and interactions. Many attempts have recently been made to reduce or mitigate the levels of PAHs in processed food products that involve optimizing the cooking methods based on numerous variables,[20] use of anti-oxidants,[21] such as tea and beer treatments[22,23] absorption of PAHs on packaging materials,[24,25] use of probiotics and lactic acid bacteria (LAB)[26] *etc.*, but the viability of these methods is raising concern. Firstly, the numerous variables identified have been considered as a part of a "Code of Practice for the Reduction of Contamination of Food with (PAH) from smoking and direct drying processes" (CAC/RCP 68/2009), but how much of the dissemination of information is being utilized by end-consumers and food producers is still unknown. Secondly, certain food products have innate anti-oxidant capacity. Antioxidants are known to reduce PAH formation in some food products. But how inbuilt food anti-oxidants behave on cooking and their reaction with PAHs has not yet been determined. Thirdly, products which themselves are known and recognized for PAH contamination contribute to the reduction of PAHs in other food products, such as meat. Meat is a widely consumed food product worldwide and is an important source of energy, protein and essential nutrients. These products are highly susceptible to the formation of PAHs due to a rich fat content and the lipophilic nature of PAHs. Thus, attempts are being made to reduce levels of PAHs in these food products during cooking and processing. As a result, the effect of different marinades such as tea,[27] oil,[28] beer[23] *etc.* on PAH reduction have been evaluated and reported successful. But surprisingly many of the products used for reduction of PAH formation are known for PAH contamination in themselves.[6,29,30] This is directly challenging our understanding of PAHs in processed food. Fourthly, various packaging materials have been shown to participate in PAH reduction in processed food.[31,32] However, packaging materials themselves can be contaminated with PAHs and phthalates.[33,34] The concern thus lies in whether it is wise to use them for mitigating PAHs while increasing the risk associated with phthalates. Fifth, the presence of certain enzymes in cells is given as a determining factor in the mechanism of PAH reduction in fermented and LAB-treated food products. Enzymes present in most of the food products might be susceptible to change due to accumulated PAHs in food products. Understanding the specific interaction between enzymes and PAHs in food, especially after cooking is a significant challenge. Lastly, storage and preservation of food through canning processes can also affect PAH concentration and toxicant stability. This especially applies to food products such as sardines, sauces and vegetables, that are transported worldwide after canning. It may be noted here that in most cases these products are smoked and cooked before canning which can increase PAH levels. Although more PAHs might accumulate in these products when steeped in preservative oils derived from pollution-affected areas,[35–37] very few studies have addressed this issue.[37]

As discussed above, processing of food itself poses a challenge in every aspect and is a continuously demanding research field. Although attempts are being made to understand more about PAHs by specific research studies and understanding of the synergistic effects through multi-residue analysis, future research is warranted.

6.4 Analytical Challenges

Detection of PAHs in processed food as stated above has been a challenging area of research for more than 60 years and has seen an evolution in analytical science and equipment development.[38] The most extensively, traditionally and commonly used analytical methods are gas chromatography flame ionization detection (GC-FID), high-performance liquid chromatography with fluorescence detection (HPLC-FLD) and gas chromatography mass spectrometry (GC-MS). These methods have often given the advantage of high sensitivity, easy handling and low-level detection. Despite the advantages offered by these analytical developments, effective determination of PAHs in processed food has always been complex due to the multifaceted nature of food and related complex matrix interferences. The determination process in general revolves around three stages of extraction, purification, clean-up and analysis through complex and often expensive analytical methods. Duedahl-Olesen,[38] Plaza-Bolanos *et al.*[39] and Zelinkova and Wenzl[40] in their reviews have raised major concerns and scientific challenges associated with PAHs in processed food. The major concerns are summarized here.

6.4.1 Extraction

The first most commonly stated and addressed concern and challenge in the extraction protocols is the co-extraction of fat from the food matrix. Liquid–liquid extraction (LLE) performed with chemicals such as dimethylformamide (DMF) or dimethylsulfoxide (DMSO), treatment with formic acid to facilitate complexation and saponification based on alcoholic potassium/

sodium hydroxide are the most commonly employed methods. Caffeine complexation is an extraction process where food is mixed in caffeine–formic acid after its dissolution in cyclohexane to form a caffeine–PAH complex. This complex is later destroyed using sodium chloride salts and re-suspending the extracts in cyclohexane.[41] This method has been explored for different food products such as coffee and edible oils.[38,42,43] This works on the principle of removing matrix interferences especially fat (*e.g.* triacylglycerols) through degradation, partition coefficients and complexation. However, the development of novel methods consuming less time and solvent are ongoing.[38] The saponification step followed by LLE although effective in removing the lipid content from matrix extracts, may also result in B(*a*)P loss to the alcohol phase upon partitioning. It has also been reported that the saponification step has a negative effect on most of the labile PAH compounds.[43] These effects have been prominently observed for milk samples.[39] Compared with LLE, matrix–solid-phase dispersion (MSPD) gives an advantage of speed and low solvent requirements but is often found to result in lower reproducibility and higher relative standard deviation (RSD) values, which are very important considerations for validating analytical methods. On the other hand, evaluation of MSPD with sulphuric acid-impregnated silica gel gave low RSD, complete fatty acid removal and adequate recovery rates.[44] However, it was observed that compounds such as benzo(*a*)anthracene, benzo(*b*)fluoranthene, benzo(*k*)fluoranthene, benzo(*a*)pyrene, dibenzo(*a*,*h*)anthracene and indeno(1,2,3-*cd*)pyrene were retained in the sorbent.[44] Solid-phase microextraction (SPME) techniques are commonly employed for liquid samples with direct extraction devices (SPME-DED) and have been evaluated in meat samples.[4] The SPME techniques in general are associated with problems of matrix effect and possible carry-over, decreased fiber efficiency and lifetime, low precision and reproducibility, and higher RSD. With SPME-DED, RSD was found to be consistent although detailed information about recovery rates has not been provided.[45] The application of this technique with solid samples has been made feasible by use of a gelatin support. Another novel technique whose application for PAHs in processed food has been in question is microwave-assisted extraction (MAE), which has indicated stability problems for a few PAHs.[39] Comparable results to LLE for Yerba mate tea have been developed using the technique stir-bar absorptive extraction with thermal desorption (SBSE-TD), as this is solvent free.[46] However, the method requires a special chromatographic system to carry out desorption of analytes. Salt sodium chloride (NaCl) used in the process (to increase the ionic strength) also demonstrated a negative effect on PAH transfer and their interaction with the sorbent bar, also known as the 'oil effect'.[46]

6.4.2 Clean-up

Several sorbents have been tested to improve the extraction of targeted PAHs. Recently, humic-acid bond silica which utilizes charge transfer and hydrophobic interactions for PAHs has been used as a new sorbent for an edible oil matrix.[47] However, the use of this material has not been evaluated for different food products. The clean-up of diluted oils performed through donor–acceptor complex chromatography (DACC) on diluted oil samples has recently been developed but is accompanied by co-elution of light PAHs (naphthalene, acenaphthylene, acenaphthene and fluorene) with fatty fractions. The use of more than 5% tetrahydrofuran (THF) further added to the co-elution problems. Higher THF concentration resulted in shifting of the PAH chromatography peak in a forward direction, thereby merging and co-eluting compounds with fatty acid peaks.[48] Another clean-up technique, gel permeation chromatography (GPC),[39] is semi-automated and is less frequently used due to moderate solvent consumption. Use of this technique for a fish matrix is not only discouraged because of solvent requirement but also due to the incorporation of a chlorinated mobile-phase solvent such as dichloromethane or chloroform which poses serious safety concerns. Hence in most cases, PAHs in a fish matrix have been extracted through Soxhlet extraction due to its high efficiency. However, Soxhlet extraction has always been known for its high solvent and time requirement, and non-selective nature. Using the same matrix, a recommendation was made to use water-miscible solvents in pressurized liquid extraction (PLE) protocols, so as to enhance the penetration into wet fatty matrices. With the PLE technique, application of saponification and sulfuric acid treatment may be accompanied by lower removal of fatty acids, decomposition of Acy and B(*a*)P PAH compounds and loss of signal within the system. Membrane-assisted solvent extraction (MASE) method often employed for beverages on the other hand has major advantages.[49] This technique was found to be relatively environment friendly and gave greater recovery rates and RSD values. Use of SPE cartridges has minimized the laborious nature of clean-up but with respect to a coffee matrix as an example, single-stage SPE resulted in clogging problems in the cartridge. The other issue which needs to be resolved before using reverse-phase or polymeric sorbents-based SPE systems is solvent use, as PAH adsorption on glass walls may occur and affect PAH solubility. It is noted that a low solvent percentage facilitates the solubilization of heavy PAHs while high solvent percentages result in the breakdown of lighter PAHs.[50] Poor recoveries of compounds (Naph, Acy and Fluo) have also been reported with silica-based SPE.[51] A recent advance in this field is the utilization of ionic liquids (IL). These extraction surfactants are non-molecular solvents possessing negligible vapor pressure, low toxicity, high thermal stability and low critical micelle concentration. These solvents have been used for testing toasted cereals with MAE and could be directly injected into HPLC-FLD systems without any clean-up.[52]

6.4.3 Analysis

Determination of non-polar PAHs has always been a matter of concern owing to their high volatility and losses. On the other hand, analysis of high molecular weight PAHs has been a challenge when GC-MS methods are employed due to their low volatility. Additionally, a challenge lies in the separation of different isomers of PAHs, such as B(*b*)F–B(*k*)F along with benzo(*j*)fluoranthene [B(*j*)F], B(*a*)A-Chry and with cyclopenta(*cd*)pyrene and Chry-triphenylene on GC-MS columns, and is another area of continuous evolution. The other disadvantage is that adequate resolution for target PAHs may not be obtained through this process due to co-elution of numerous other compounds, especially when the overlapping or interfering compounds have isobaric ions.[53,54] This problem is critically faced by the above three compound groups along with DB(*ah*)A and IP, and DB(*ah*)A and dibenzo[*a*,*c*]anthracene [DB(*ac*)A]. Dibenzopyrenes are among the heavy PAHs that interact with GC columns (typical 5 ms) and broaden the peak detection with compromised sensitivity. The head-space (HS) SPME technique has recently been evaluated for PAHs in milk[55] and offers the advantages of minimal sample handling and reduced solvent consumption. However, it was not found suitable for heavy PAHs and gave higher RSD values, which again are a concern. With tea infusions, numerous fibers have been tried using the HS-SPME method, but insufficient film coating of the fibers resulted in overloading problems.[46] Also, it may be noted that the effect of sample homogenization with HS-SPME has yet not been illustrated due to a lack of information on overall performance characteristics.

HPLC-FLD thus is the most common technique employed. However, the technique exhibits various critical disadvantages such as selectivity problems between parent and alkylated PAHs (an impurity when only parent PAH is of concern). Unsubstituted PAHs exhibit a similar fluorescence response, although certain isotopically labeled compounds cannot be used due to their similarity with parent PAHs. Furthermore, monitoring of PAH compounds with no native fluorescence is a drawback of this technique thereby making GC-MS (/MS) comparatively superior. FLD in combination with ultra-high-pressure liquid chromatography (UHPLC) reduces time, improves chromatographic resolution and discriminates co-eluted interferences within a food matrix.

6.4.4 Less Explored Techniques

- Supercritical fluid extraction (SFE) is a technique widely used for food applications.[56,57] Extraction of PAHs using this technique using a C18 column has been studied. However, the technique is underutilized despite promising results.[58,59]
- Determination of PAHs in processed food products through LC-MS techniques has also not gained widespread use. This might be due to the decreased ionization efficiencies, troublesome presence of water vapor and decreased method sensitivity, which has been overcome in more recent applications.[60]
- Ionization techniques such as atmospheric-pressure chemical ionization source (APCI) and atmospheric pressure photo-ionization (APPI)

have dealt with parent- and hydroxyl-PAHs in processed food, even though they are designed for non-polar compounds.[61] A single study was found where an LC-MS-ESI (electrospray ionization) technique was used for hydroxyl-PAH determination.[62] However, the requirement for ionization and derivatization steps has been highlighted as complex techniques require highly trained personnel. It should also be noted that several studies using LC–GC or LC–LC–GC techniques exist for alkylated PAHs in mineral oil to address contamination, but in the context of food studies are limited. These techniques should be exploited due to their innovative nature and reduced sample handling. The feasibility of APPI sources for non-EPA PAHs should also be evaluated. Comparison of these techniques with conventional LC-FLD may further enhance knowledge about simultaneous identification and confirmation of PAH moieties.

- Similarly, UHPLC instruments have been widely exploited for trace-level detection of organic compounds other than PAHs[63] and provide research scope for future studies. Working on these analytical instruments and aspects may further aid increased resolution and selectivity in very complex matrices.
- Similarly, GC-APLI (atmospheric pressure laser ionization) MS detection studies have not been explored for food applications, despite their wide application in environmental samples.[40]
- Specific GC columns have been developed for analyzing dibenzopyrenes. However, limited studies focus on the use of these columns for food have been reported.[39,64]
- Some other rarely applied techniques are programmed temperature vaporization (PTV), cold on-column injection and large volume injection (LVI) which possess advantages of improving peak shape and early elution of PAHs, increased sensitivity and diminishing discrimination of dibenzopyrenes.[54] Similarly, single quadrupole analyzers (Q) using single-ion monitoring mode (SIM) have been extensively studied but ion trap (IT) and triple quadrupole (QqQ) are the least developed techniques[65]
- Recently, the focus has been shifted to the incorporation of nano-particles for detection and extraction of PAHs[66,67] However, further in-depth studies on their feasibility and robustness are warranted. For example, Li *et al.*[67] extracted PAH from food samples using hybrid microspheres made of a new core–shell nanostructure and magnetic covalent organic framework [Fe_3O_4@COF-(TpBD)]. The analysis was then performed using a HPLC photodiode array detector (HPLC-DAD) technique. The results demonstrated the fast adsorption ability of these microspheres. In another study, Wang *et al.*[66] studied the adsorption and desorption of PAH on g-C_3N_4/Fe_3O_4 (graphitic carbon nitride/ferric oxide) nanocomposites to facilitate simple magnetic solid-phase extraction (MSPE).

6.4.5 Other Challenges

- With respect to different food matrices, very few certified reference materials (CRMs) for PAHs are available. Proficiency testing in such cases may prove suitable for quality control.[40] The lack of CRMs has also been highlighted for spirits and whisky in particular[68] and deodorizer distillates.[69]
- Very few studies have focused on the effect of preparation methods such as autoclaving and freeze drying on the development of CRMs.[70] The underlying challenges in this field are converting incurred material (starting material with close resemblance to real samples) to match real processed food products with minimum manipulation, reducing heterogeneity and increasing the stability of targeted compounds in them to facilitate storage and shipping.
- A concern also arises when samples are stored for an extended time period before analysis as this may result in or favor interactions between PAHs and the food matrix.[71]
- A challenge also lies in determining a comparison between the limit of quantification (LOQ) obtained for a particular food product by GC-MS (/MS) and LC-FLD as reported LOQs are absent in many studies.[72,73]
- A concern highlighted in most of the studies on tea has reported less solubility of PAHs in water resulting in lower detection.[74]

6.5 Challenges in Assessing Health Risk Using PAH Data

The risk assessment associated with dietary intake of PAHs from processed food comprises two steps. First the analysis of PAH contamination in food and secondly the application of this information to food consumption, in order to calculate an overall exposure risk.

6.5.1 Concerns Regarding the Reported PAH Data

The interpretation of data used in different studies on levels of PAHs present in processed food to assess risk associated with their consumption is challenging. This often prevents complete dietary risk assessments. First of all, few studies have focussed on evaluating risk from a complete dietary profile as most studies are product-specific. Second, reviewing complete data based on these scattered food products and analyses to give meaningful results is a complex process. Furthermore, in some studies, the concentration of two or more PAHs is reported as a single value due to merging of peaks, which makes comparison of individual PAH concentrations difficult with other studies. Choosing a toxicity equivalency factor (TEF) for the combined peaks derived from chromatographic techniques is problematic. In such cases, if the TEFs of two compounds are added, assuming that their effect is additive, then the results deviate and can contradict in the case where the two compounds share an antagonistic nature as demonstrated by the dose response animal studies. However, non-addition, will potentially underestimate the risk faced by a population. Third, the PAH concentration calculated in different studies across the world are either estimated on a dry weight basis or on wet weight basis, thereby making comparison of the two techniques problematic. Thus, PAH levels in a single food product are often reported in these two ways and their relative conversion factors are lacking.[75] As a result, choosing studies or ignoring the moisture factor (which plays an important role in PAH formation and accumulation in food during processing) results in a biased assessment of risk. Another analytical factor which directly affects the assessment of risk is the unit measurement in which the final concentrations are reported. For example, for coffee the PAH concentration in studies are reported as $\mu g\ L^{-1}$ in a few studies and as $mg\ kg^{-1}$ in other studies.[75] This makes comparison of these studies on a single platform difficult.[75] It may be noted that homogeneity of the collected data is a prime requirement for critical analysis of risk associated with exposure to contaminants. The most important factor affecting this parameter is the type of instrument used and the analytical conditions.

6.5.2 Lack of Consumption Data

Consumption data is collected from individuals through surveys and questionnaires addressing their consumption patterns. This is then used for calculating the toxicant consumption data. This data is often collected from relatively small population groups (representative sample) rather than regional or national level data collection. As a result, the huge variability existing throughout a region or on a national scale with respect to the cooking styles, food patterns, nutritional habits, available food quantities *etc.*, are not factored into the data, leading to biased risk assessment. Bansal and Kim[76] highlighted that relatively very few studies have dealt with understanding the relationship between PAH consumption and consistent human exposure. A second constraint lies in not having the consumption data for a particular food group. In this case, the assessment of risk cannot be determined and as such data has to be eliminated.[75] Furthermore, food products are not usually consumed solely as a single component within a meal. These complex matrix interactions cannot be determined. At times they are transformed into forms which combine ingredients from one or more food categories. In this case, estimation of risk cannot be performed on the basis of one food item. Additionally, the formula for risk assessment is lacking in consideration of different influencing factors such as poverty line, malnutrition, economic status, food preferences and income, resulting in ambiguity and difficulty in the interpretation of results. Interpretation of data especially in cases where samples are collected from commercialized outlets such as restaurants and fast-food chains becomes highly problematic as the details related to cooking are not readily available and the number of samples analyzed are often limited. Domingo and Nadal[77] stated in their recent review that the methodologies used in such surveys are often different worldwide, focusing on specific food products, different PAH analysis, different age groups *etc.* and thus comparison of these studies cannot be performed. Uncertainties arise from duplicate diets and basket studies when performing total diet surveys, as background information and preparation methods at a household level are not available. Furthermore, in such studies it is considered that the consumer has eaten all of the food product, however, in reality the same consumer may not necessarily have the same food all the time or throughout the period accounted for in calculations of overall exposure to the contaminant.

6.5.3 Approach of Risk Assessment

Duedahl-Olesen *et al.*[38] and Benford and DiNovi[10] in their reviews stated that the TEF approach to dietary risk assessment is not appropriate owing to limited availability of data on mode of actions of different PAHs in the causation of cancer and other adverse human health effects. Also, different studies[78,79] have reported varying TEF values, with most compounds still lacking information about their oral doses and response in animal studies.[10] Secondly, toxicity equivalence factors (TEFs) and margins of exposure (MOEs) are derived from the impact and responses found in animal studies for the different coal tar mixtures. This approach seems relevant as applications of pure compounds in combination does not match their theoretical additive results and shows repressed responses. This drawback is not evident on feeding coal tar mixtures to study groups. However, in reality the charred material and PAH rich textural part of food products often differs from these coal tar mixtures. A range of variability exists ranging from dry soot-charred carbon to oily black liquid, thus making all assessments based on coal tar mixtures irrelevant. The challenge thus lies in matching the reality and actual dosages for real risk assessments of dietary intake of PAHs present in processed food.

6.6 Research Gaps

The research gaps identified from information collected from published studies focusing on PAHs in processed food are as follows.

6.6.1 Status of Research Studies

Occurrence of PAHs have been reported in many food products such as milk, cheese, fish, bread, meat, grains, chocolate *etc.*[6,9] Despite several studies targeting a variety of food products, there is a lack of studies reporting PAHs in spices,[80] butter[81] and many other products. Similar to fish and meat, smoked cheese is a fatty matrix product and monitoring of PAHs in this food category is a topic of interest.[39] But very few studies have addressed these food products. The European Union (EU) has also highlighted scarcity of information available on the presence of PAHs in bread.[76] Singh and coworkers[6] in their review highlighted the use of various novel techniques to reduce PAH levels, such as ohmic heating, infrared cooking *etc.* for food preparation. However, very few studies have focused on these processes with respect to PAH formation.

Domingo and Nadal[77] in their comprehensive review stated that the scientific research papers published on PAHs in processed foods were fewer for countries like Australia, Canada, Germany, or Japan, despite the concerns in the scientific world about risk associated with exposure to environmental contaminants and food toxicants. Italy, Spain, the United Kingdom and the Netherlands have further reported dietary intake of PAHs while limited information for the same was observed for other countries such as Estonia, Azerbaijan, France, Sweden and Greece. In India, the studies on PAHs have been focused only on specific food products instead of complete dietary assessment. Considering the limited spread of studies on PAHs in processed food products from different parts of the world, it is difficult to state much about the temporal trends of these toxic PAH compounds. The reviews have also highlighted the temporal trend for PAH concentrations in the USA during the past 10 years. The levels of these compounds have remained the same during the period, but to date the associated dietary risk for PAHs are not available. Similarly, the risk assessments are not present for any of the African countries.

6.6.2 Mechanism for PAH Formation in Processed Food

Although the main reason for PAH formation in a food matrix has been reported as pyrolysis of organic matter, three main mechanisms have been identified with respect to food. The first mechanism involves pyrolysis of major organic components of a food product *i.e.* fat, carbohydrates and protein at temperatures above 200 °C followed by PAH formation at higher temperatures ranging between 500 and 900 °C. As for the second mechanism, PAH formation can take place from direct dripping of fat from food products onto a heat source and the accumulation of the compounds formed on food after contact with contaminated/polluted smoke. Lastly, the burning of fuel used for cooking purposes especially charcoal may produce PAHs, which may then accumulate in food products on exposure to flames during the cooking process.[82] Though the three mechanisms for formation of PAHs in food have been described, the exact mechanisms are not precisely known.

6.6.3 Types of PAHs Analyzed

The most commonly studied PAHs in food are those prioritized by the US-EPA and the EU. However, recently many studies have found alkylated, nitro and oxygenated PAHs in food products.[83] Oxygenated PAHs (*O*-PAHs) are transformed into PAH products possessing hydroxyl, carbonyl and carboxyl functional groups and are known for their direct mutagenic properties.[84] Also, alkylated PAHs have been reported to be more toxic than parent PAHs. Sun *et al.*[2] in their review highlighted the need for conducting further studies on these PAHs to gather data on their occurrence in food products, their formation mechanism and their toxicity levels so as to establish proper guidelines and regulations for the same. Technological development for the detection of these derivatives also forms part of a major research gap. It would be interesting to understand how these PAHs affect cell and tissue systems and especially reproduction status. For this, the assessment of their transferrable rate between cells and tissues would prove beneficial.[85,86]

6.6.4 Diagnostic Ratios

Diagnostic ratios have always been used to identify the sources of PAHs found in any given matrix. These ratios are extensively utilized in environmental monitoring programmes. These ratios are well established for PAHs in air and for determining sources of PAHs in residential and commercial kitchens where cooking of food is performed.[87] However, studies on PAHs in processed food have rarely utilized these diagnostic ratios.[88] Diagnostic ratios may aid in differentiating the cooking method from other PAH sources, thus discriminating between the nature of food products and the cooking styles used to prepare them. Choosing the right set of ratios and developing values specific for each cooking style and for respective food products may prove beneficial. Also, these diagnostic ratios may help in identifying the exact contamination source if monitored throughout the supply chain for products like oil, milk powders *etc.*

6.6.5 Lack of Information About Processing

Another major concern highlighted by Domingo[16] is a lack of information on the effect of cooking activities on levels of food contaminants. That work also stated that the studies conducted for food analysis mostly focuses on unprocessed/raw products, ignoring the fact that the modifications take place as physicochemical reactions and affect nutritional qualities of food during cooking or processing. However, the impact of pre-harvest stages or unit operations on post-harvest PAH formation and accumulation by a food product is worth determining. It should also be noted in this case that the studies that focus on commercial sample analysis do not have most of the background information about preparation techniques. This results in an incomplete picture of the mechanism involved in PAH formation. Bansal and Kim[76] highlighted this constraint while analysing details of the presence of PAHs in chocolate. In their review, they stated that it is difficult to correlate PAH concentrations with the labelled cocoa contents as labels of chocolate did not possess adequate information. Sun *et al.*[2] also stated in their study that PAH concentration in olive oil is a matter of concern owing to its importance in consumer consumption. The labels used for olive oil packaging do not provide any information about PAH content and related adulteration issues. However, adulteration and mislabelling should be avoided keeping in view consumer health concerns. In future, it would be worth looking at food labelling practices and symbols could be provided showing the levels of PAHs in order to raise consumer awareness. Consumer interest and preferences for such labels is again an area of study.

6.6.6 Fate of By-products and Mitigation Strategies

In a recent study conducted by Shi *et al.*,[69] the PAH content of the deodorizer distillate, obtained after edible oil refining, which is often used in nutraceutical and vegetable cream formulations, cosmetics and food supplements, was analyzed. Refining of oil is often considered as a strategy to reduce PAH contamination of crude oils, but the distillate obtained after the refining process was found to be rich in PAHs, which again adds to dietary risk when incorporated into food supplements. Hence, it becomes necessary to critically evaluate the reduction and mitigation strategies and find improved ways to prevent PAH contamination. The other technique targeted nowadays to reduce PAH contamination in edible oils is treatment with activated carbon. But the technique has been of concern owing to its ability to alter the bioactive compounds and oxidative stability, consequently affecting product preferences of consumers.[28,89]

Mitigation of PAH contamination in smoked food products mainly involves application of anti-oxidants and phenols. However, there are concerns with respect to the levels for application of these phenolic and anti-oxidant compounds due to their cytotoxicity and mutagenicity.[90] In another study, Farhadian *et al.*[91] reported PAH reduction in beef satay using lemon juice (acidic marinade) when cooking oil is used as an ingredient in the marinade. Similarly, investigation of compounds present in tea infusions which can increase or decrease PAH content in chicken wings needs to be undertaken and the mechanism for reduction needs to be identified.[27] This reflects that PAH reduction strategies need a broader approach.

Utilization of γ-irradiation and microwave techniques for PAH reduction in food products such as cereal grains have shown promising results. However, Khalil *et al.*[92] has highlighted the need of carrying out such investigations with wider dose ranges on diverse food products before providing a firm conclusion. Similarly, Kiralan *et al.*[93] defined the need for development of a microwave system with tunnel technology which may be used at an industrial level to reduce PAH contamination in olive oil through microwave preheating. They have also stated the need to understand the effect of variables during such applications. These variables include dielectric properties of the food product, power/temperature applications, moisture content of olive pomace, surface moisture uniformity and scaling-up processes. Parker *et al.*[94] have also highlighted that the utilization of ultraviolet light and oxygen for downstream processing of smoked food products must also consider the fact that little information is available at present about the toxicity of radiation and the resulting derivatives, as in the future these derivatives might prove more toxic than PAHs. Hence selection of a defined reduction technique needs a broader view as well as further research.

6.6.7 Regulatory and Legislative Gaps

Although different guidelines for PAHs in food have been established by the EU, European Food Safety Authority (EFSA), National Food Safety Standards (China), National Health Surveillance Agency (Brazil) *etc.*, the guidelines are still lacking in one respect or another owing to the diversity of food products, cooking methods and other complexities involved in food production and preparation. Also, guidelines based on 15 genotoxic PAHs and 8 carcinogenic PAHs (8PAH) needs further refinement and definition. It must be noted that most studies have detected an abundance of light PAHs in a variety of food products. But light PAHs are excluded from the list of 8PAH defined by EFSA. Also, oral toxicity is still a matter of question.[76] It should also be noted that EFSA has recognized that the guidelines on PAHs in processed food lack information on food categories such as cereals and grains, spices and herbs, beverages (herbal infusions, tea, coffee), which are part of the daily food intake and often ingested at large amounts.[39] Vegetables form an important food category in the human diet, especially for population groups who prefer either a vegetarian or vegan diet. Despite the high nutritive value and consumption of vegetables, legislative guidelines have not been formulated for this food category. However, much work has been focussed on reporting and mitigating levels of PAH in predominantly animal-based food products. Thus, importance must be given to a vegetarian or vegan diet, and the risk faced by people purely following vegetarian or vegan diets must be addressed and guidelines for these categories should be given to substantially help consumers. The occurrence of benzo[*c*]fluorene (BcF) in food has been highlighted as a concern by EFSA but has not been included in EPA and EU lists for PAHs. Also, a number of new PAH derivatives have been found and reported in food[86] and due to their higher toxic nature than parent PAHs the immediate need lies in incorporating them into statutory regulations. This further highlights the need of having robust data on PAH levels in processed food, extensive sampling, reduction in errors and development of definitive detection methodologies.[39]

6.7 Research Recommendations

Based on the challenges and gaps in research on PAH levels in processed food identified, some research recommendations can be made as follows.

6.7.1 Standardization of Food Products and Processes

Development of standardized cooking methods for food products at a global and national scale that maintains PAH content at safe limits is required. Any deviation from these standard procedures will help in estimating the risk faced by consumers, and the factors which lead to these deviations should be determined.

6.7.2 Studying Hidden Influential Factors

Some factors that may prove beneficial if addressed include the following:

(a) The relationship between oxygen concentration and its accessibility/supply to food components and PAHs.
(b) The effect of air circulation on PAH formation in smoking chambers.
(c) Addressing changes in food structure and quality upon cooking and processing
(d) Developing tools to analyze non-log-normal distribution data and worst-case scenarios as in reality there are many deviations
(e) Understanding the correlation between food composition and PAH occurrence
(f) Addressing the effect of minimal processing and high-pressure processing techniques[95] on quality and nutritional values of food products.
(g) Predicting the rate at which PAHs transfer from different food products their and interactions with different human cell lines
(h) Studying the effect of salt and other ingredients (osmosis) applied during cooking on PAH formation and solubility in food.
(i) The effect of photo-oxidation and light treatments of food before and after cooking.
(j) Determining if sulfoxides or sulfones and pro-oxygens formed in food during cooking can be used to sense PAH concentration.
(k) Determining the mechanisms of PAH reduction strategies.

6.7.3 Designing Cooking Equipment/Tools

Researchers who specialize in different aspects of food analysis and development, *e.g.* toxicologists, food engineers, technologists, may come together to design a suitable grill or smoking chamber which may ensure proper and uniform cooking of food with minimized PAH formation and accumulation.[96] Thus instrument/machine design for reduced PAH formation/contamination is an interesting area of research.

6.7.4 Development of Mathematical Models

Attempts must be made to combine other parameters of food with PAHs. For example, building an equation, correlation and mathematical model based on color,[97] antioxidant content, pro-oxygen compound analysis, pH, temperature imaging of food to determine a theoretical number of PAHs formed in a food product may be explored.

6.7.5 Developing Initial Screening and Rapid Detection Tools for Samples

Separating the number of food samples at an initial level by screening them as non-carcinogenic and carcinogenic will save time and resources. This will reduce the burden on costly analytical techniques. However, the method or tools used should be effective enough to detect even low levels of PAHs *i.e.* near the defined limits in food regulations which may pose human health threats. Attempts have been made in this direction by incorporating fluorescence spectroscopy techniques into the determination of PAH levels in edible oils.[98] However, at present this technique is at an early development stage and facing challenges such as interference from tocopherols (a fat-soluble vitamin), and other pigments, inner filter effects upon fluorophore–environment interaction, limited PAH monitoring (restricts use for those PAHs that lack fluorescence properties), multi-variate calibration and statistical tools for data analysis (spectral data). Additionally, different sensing techniques are being developed. It would be wise to evaluate their efficiency for quantifying PAHs in processed food products as a rapid detection tool. Attempts must be made to incorporate more environmentally friendly solvents and techniques for detection of PAHs.

6.7.6 Analyzing Risk–Benefit and Continuous Scenario-Based Monitoring

Further research is required to analyze the risk–benefits of food products on a common scale, as in the European Funded Project BRAFO (benefit–risk analysis of foods).[99] In this case, incorporating inter-laboratory assessments for the same sample sets may prove beneficial. The project will account for the variability in analytical equipment and methodology and will aid in developing a correlation between them. Also, it must be kept in mind that end-consumers should be the main beneficiaries so that consumers are helped to understand the level of risk they are facing from eating PAHs in processed food. The results of the project should be expressed in convenient units that consumers can relate to and understand with ease. Research must also consider the variation of PAH levels in defined products when prepared by consumers in the home environment.[100] Lastly a discussion platform, project results and accessible database should be created at the global scale to address PAH levels in processed food products. With an increase in identification of PAH derivatives, discussion regarding the modification of guidelines will require a combined effort of researchers, legislators and food manufacturers to mitigate the risk of PAHs in food products.

Acknowledgement

The first author L.S. is thankful for the scholarship provided by NIFTEM for carrying out the research work.

References

1. B. O. Silva, O. T. Adetunde, T. O. Oluseyi, K. O. Olayinka and B. I. Alo, *Afr. J. Food Sci.*, 2011, **5**(7), 384.
2. Y. Sun, S. Wu and G. Gong, *Trends Food Sci. Technol.*, 2019, **83**, 86.
3. I. J. Keyte, R. M. Harrison and G. Lammel, *Chem. Soc. Rev.*, 2013, **42**, 9333.
4. T. Agarwal, P. S. Khillare, V. Shridhar and S. Ray, *J. Hazard. Mater.*, 2009, **163**, 1033.
5. C. Sun, J. Zhang, Q. Ma, Y. Chen and H. Ju, *Environ. Geochem. Health*, 2017, **39**, 63.
6. L. Singh, J. G. Varshney and T. Agarwal, *Food Chem.*, 2016, **199**, 768.
7. L. Becker, D. P. Glavin and J. L. Bada, *Geochim. Cosmochim. Acta*, 1997, **61**(2), 475.
8. F. J. Lovas, R. J. McMahon, J. Grabow, M. Schnell, J. Mack, L. T. Scott and R. L. Kuczkowski, *J. Am. Chem. Soc.*, 2005, **127**, 4345.
9. L. Singh and T. Agarwal, *Trends Food Sci. Technol.*, 2018, **79**, 160.
10. D. Benford, M. DiNovi and R. W. Setzer, *Food Chem. Toxicol.*, 2010, **48**, S42.
11. G. Purcaro, S. Moret and L. S. Conte, *Talanta*, 2013, **105**, 292.
12. I. W. H. Jarvis, K. Dreij, A. Mattsson, B. Jernstrom and U. Stenius, *Toxicology*, 2014, **321**, 27.
13. S. Pavanello, A. C. Pesatori, L. Dioni, M. Hoxha, V. Bollati, E. Siwinska, D. Mielzynska, C. Bolognesi, P. A. Bertazzi and A. Baccarelli, *Carcinogenesis*, 2010, **31**(2), 216.
14. V. Bollati, A. Baccarelli, L. Hou, M. Bonzini, S. Fustinoni, D. Cavallo, H. Byun, J. Jiang, B. Marinelli, A. C. Pesatori, P. A. Bertazzi and A. S. Yang, *Cancer Res.*, 2007, **67**(3), 876.
15. D. N. Das and S. K. Bhutia, *Curr. Opin. Food Sci.*, 2018, **24**, 16.
16. J. L. Domingo, *Crit. Rev. Food Sci. Nutr.*, 2011, **51**, 29.

17. I. C. Shaw, *Int. J. Food Sci. Technol.*, 2014, **49**, 2149.
18. J. L. Domingo, *Food Res. Int.*, 2014, **56**, 159.
19. W. Nie, K. Cai, Y. Li, S. Zhang, Y. Wang, J. Guo, C. Chen and B. Xu, *Molecules*, 2018, **23**, 3377.
20. E. Ledesma, M. Rendueles and M. Díaz, *Food Control*, 2016, **60**, 64.
21. G. Gong, X. Zhao and S. Wu, *Food Control*, 2017, **87**, 117.
22. N. U. Benson, O. H. Fred-Ahmadu, J. A. O. Olugbuyiro, W. U. Anake, A. E. Adedapo and A. A. Olajire, *J. Food Compos. Anal.*, 2018, **66**, 13.
23. O. Viegas, I. Yebra-Pimentel, E. Martínez-Carballo, J. Simal-Gandara and I. M. P. L. V. O. Ferreira, *J. Agric. Food Chem.*, 2014, **62**, 2638.
24. K. Park, H. Pyo, W. Kim and K. S. Yoon, *Meat Sci.*, 2017, **129**, 1.
25. K. Kuźmicz and A. Ciemniak, *J. Environ. Sci. Health, Part B*, 2018, **53**(1), 1.
26. E. Bartkiene, V. Bartkevics, E. Mozuriene, V. Krungleviciute, A. Novoslavskij, A. Santini, I. Rozentale, G. Juodeikiene and D. Cizeikiene, *Food Control*, 2017, **71**, 285.
27. C. Wang, Y. Xie, J. Qi, Y. Yu, Y. Bai, C. Dai, C. Li, X. Xu and G. Zhou, *Food Control*, 2018, **93**, 325.
28. L. Shi, L. Zheng, Q. Jin and X. Wang, *Eur. J. Lipid Sci. Technol.*, 2017, **119**, 1700150.
29. L. D. Olesen, M. Navaratnam, J. Jewula and H. Jensen, *Polycyclic Aromat. Compd.*, 2015, **35**(1), 74.
30. M. Ciecierska and M. W. Obiedzinski, *J. Environ. Sci. Health, Part B*, 2013, **48**(11), 993.
31. J. Chen and S. Chen, *Food Chem.*, 2005, **90**, 461.
32. J. Semanova, B. Sklarsova, P. Simon and P. Simko, *Food Chem.*, 2016, **201**, 1.
33. S. Li, H. Ni and H. Zeng, *Sci. Total Environ.*, 2017, **609**, 1126.
34. R. H. Dong, H. Zhang, M. R. Zhang, J. S. Chen, M. Wu, S. G. Li and B. Chen, *Biomed. Environ. Sci.*, 2017, **30**, 727.
35. N. J. Novakov, Z. A. Mihaljev, B. D. Kartalovic, B. J. Blagojevic, J. M. Petrovic, M. A. Cirkovic and D. R. Rogan, *Food Addit. Contam., Part B*, 2017, **10**(3), 208.
36. S. Moret, G. Purcaro and L. S. Conte, *Eur. J. Lipid Sci. Technol.*, 2005, **107**, 488.
37. N. U. Benson, W. U. Anake, A. E. Adedapo, O. H. Fred-Ahmadu and K. P. Eke, *Food Compos. Anal.*, 2016, **57**, 109.
38. L. Duedahl-Olesen, in *Persistent Organic Pollutants and Toxic Metals in Foods*, ed. M. Rose and A. Fernandes, Woodhead Publishing, UK, 1st edn, 2013, vol. 13, p. 308.
39. P. Plaza-Bolanos, A. G. Frenich and J. L. M. Vidal, *J. Chromatogr. A*, 2010, **1217**, 6303.
40. Z. Zelinkova and T. Wenzl, *Polycyclic Aromat. Compd.*, 2015, **35**(2–4), 248.
41. K. Tamakawa, in *Handbook of Food Analysis Volume 2: Residues and Other Food Component Analysis*, ed. L. M. L. Nollet, Marcel Dekker Inc., USA, 1st edn, 2004, vol. 38, p. 1449.

42. A. N. Sagredos, *Dtsch. Lebensm.-Rundsch.*, 1979, **75**(11), 350.
43. S. Moret and L. S. Conte, *J. Chromatogr. A*, 2000, **882**, 245.
44. L. Pensado, M. C. Casais, M. C. Mejuto and R. Cela, *J. Chromatogr. A*, 2005, **1077**(2), 103.
45. D. Martin and J. Ruiz, *Talanta*, 2007, **71**, 751.
46. V. Gomes-Zuin, L. Montero, C. Bauer and P. Popp, *J. Chromatogr. A*, 2005, **1091**, 2.
47. D. Luo, Q. Yu, H. Yin and Y. Feng, *Anal. Chim. Acta*, 2007, **588**, 261.
48. A. Barranco, R. M. Alonso-Salces, E. Corta, L. A. Berrueta, B. Gallo, F. Vicente and M. Sarobe, *Food Chem.*, 2004, **86**, 465.
49. V. Gomes Zuin, M. Schellin, L. Montero, J. H. Yariwake, F. Augusto and P. Popp, *J. Chromatogr. A*, 2006, **1114**, 180.
50. J. K. Houessou, C. Benac, C. Delteil and V. R. Camel, *J. Agric. Food Chem.*, 2005, **53**, 871.
51. M. Suchanová, J. Hajšlová, M. Tomaniová, V. Kocourek and L. Babička, *J. Sci. Food Agric.*, 2008, **88**, 1307.
52. M. German-Hernandez, V. Pino, J. L. Anderson and A. M. Afonso, *J. Chromatogr. A*, 2012, **1227**, 29.
53. D. L. Poster, M. M. Schantz, L. C. Sander and S. A. Wise, *Anal. Bioanal. Chem.*, 2006, **386**, 859.
54. J. A. Gómez-Ruiz and T. Wenzl, *Anal. Bioanal. Chem.*, 2009, **393**, 1697.
55. N. Aguinaga, N. Campillo, P. Vinas and M. Hernández-Córdoba, *Anal. Bioanal. Chem.*, 2008, **391**, 753.
56. J. L. Bernal, M. T. Martin and L. Toribio, *J. Chromatogr. A*, 2013, **1313**, 24.
57. A. Tarafder, *Trends Anal. Chem.*, 2016, **81**, 3.
58. M. Y. Ali and R. B. Cole, *Anal. Bioanal. Chem.*, 2002, **374**, 923.
59. M. Y. Ali and R. B. Cole, *J. Agric. Food Chem.*, 2001, **49**, 4192.
60. D. B. Robb, T. R. Covey and A. P. Bruins, *Anal. Chem.*, 2000, **72**, 3653.
61. G. Lien, C. Chen and C. Wu, *Rapid Commun. Mass Spectrom.*, 2007, **21**, 3694.
62. L. Rey-Salgueiro, E. Martínez-Carballo, M. S. García-Falcón, C. González-Barreiro and J. Simal Gándara, *Food Chem.*, 2009, **115**, 814.
63. H. G. J. Mol, P. Plaza-Bolanos, P. Zomer, T. C. de Rijk, A. A. M. Stolker and P. P. J. Mulder, *Anal. Chem.*, 2008, **80**, 9450.
64. J. Djinovic, A. Popovic and W. Jira, *Meat Sci.*, 2008, **80**, 449.
65. J. Nácher-Mestre, R. Serrano, T. Portolés-Nicolau, F. Hernández, L. Benedito-Palos and J. Pérez-Sánchez, *Rapid Commun. Mass Spectrom.*, 2009, **23**, 2075.
66. M. Wang, S. Cui, X. Yang and W. Bi, *Talanta*, 2015, **132**, 922.
67. N. Li, D. Wu, N. Hu, G. Fan, X. Li, J. Sun, X. Chen, Y. Suo, G. Li and Y. Wu, *J. Agric. Food Chem.*, 2018, **66**(13), 3572.
68. J. I. Cacho, N. Campillo, P. Vinas and M. Hernandez-Cordoba, *Food Chem.*, 2016, **190**, 324.
69. L. Shi, L. Zheng, R. Liu, M. Chang, J. Huang, Q. Jin and X. Wang, *Food Chem.*, 2019, **275**, 206.

70. J. F. Huertas-Pérez, L. R. Bordajandi, B. Sejerøe-Olsen, H. Emteborg, A. Baù, H. Schimmel and M. Dabrio, *Anal. Bioanal. Chem.*, 2015, **407**, 3069.
71. I. Yebra-Pimentel, R. Fernández-González, E. Martínez-Carballo and J. Simal-Gándara, *Crit. Rev. Food Sci. Nutr.*, 2013, **50**, 1383.
72. B. Veyrand, A. Brosseaud, L. Sarcher, V. Varlet, F. Monteau, P. Marchand, F. Andre and B. Le Bizec, *J. Chromatogr. A*, 2007, **1149**, 333.
73. E. Ballesteros, A. García Sánchez and N. Ramos Martos, *J. Chromatogr. A*, 2006, **1111**, 89.
74. J. Lee, J. Jeong, S. Park and K. Lee, *Food Control*, 2018, **92**, 286.
75. L. Singh and T. Agarwal, *Chemosphere*, 2018, **202**, 366.
76. V. Bansal and K. H. Kim, *Environ. Int.*, 2015, **84**, 26.
77. J. L. Domingo and M. Nadal, *Food Chem. Toxicol.*, 2015, **86**, 144.
78. I. C. T. Nisbet and P. K. Lagoy, *Regul. Toxicol. Pharmacol.*, 1992, **16**, 290.
79. D. Delistraty, *Toxicological and Environmental Chemistry*, 2008, vol. 64, p. 81.
80. I. Rozentale, A. Y. Lun, D. Zacs and V. Bartkevics, *Food Control*, 2018, **83**, 45.
81. S. Amirdivani, N. Khorshidian, M. G. Dana, R. Mohammadi, A. M. Mortazavian, S. L. Q. De Souza, H. B. Rocha and R. Raices, *Dairy Technol.*, 2019, **72**(1), 120.
82. H. Alomirah, S. Al-Zenki, S. Al-Hooti, S. Zaghloul, W. Sawaya, N. Ahmed and K. Kannan, *Food Control*, 2011, **22**, 2028.
83. G. Li, S. Wu, J. Zeng, L. Wang and W. Yu, *Food Chem.*, 2016, **209**, 123.
84. G. Gong, X. Zhao and S. Wu, *Food Control*, 2018, **87**, 117.
85. A. R. Jafarabadi, A. R. Bakhtiari, Z. Yaghoobi, C. K. Yap, M. Maisano and T. Cappello, *Chemosphere*, 2019, **215**, 835.
86. M. A. Hossain and S. M. Salehuddin, *Arabian J. Chem.*, 2012, **5**, 391.
87. C. M. A. Iwegbue, E. O. C. Iteku-Atata, E. W. Odali, F. E. Egobueze, G. O. Tesi, G. E. Nwajei and B. S. Martincigh, *Exposure Health*, 2018, DOI: 10.1007/s12403-018-0276-z.
88. S. Chawda, A. Tarafdar, A. Sinha and B. K. Mishra, *Polycyclic Aromat. Compd.*, 2017, 1–12.
89. B. R. Shin, S. Yang, H. Song, M. Chung and Y. Kim, *Food Sci. Biotechnol.*, 2015, **24**(6), 2017.
90. J. Gomez-Estaca, M. C. Gomez-Guillen, P. Montero, P. Sopelana and M. D. Guillen, *LWT--Food Sci. Technol.*, 2011, **44**, 1517.
91. A. Farhadian, S. Jinap, A. Faridah and I. S. M. Zaidul, *Food Control*, 2012, **28**, 420.
92. A. Khalil, M. Albachir and A. Odeh, *Polycyclic Aromat. Compd.*, 2016, **36**(5), 873.
93. S. S. Kiralan, F. Erdogdu and A. Tekin, *Eur. J. Lipid Sci. Technol.*, 2017, **119**, 1600241.
94. J. K. Parker, S. Lignou, K. Shankland, P. Kurwie, H. D. Griffiths and D. A. Baines, *J. Agric. Food Chem.*, 2018, **66**(10), 2449.

95. K. S. Bravo, R. Ramírez, R. Durst, Z. J. Escobedo-Avellaneda, J. Welti-Chanes, P. D. Sanz and J. A. Torres, *J. Food Sci.*, 2012, **71**, R1.
96. J. Y. Yi, H. J. Kim and M. Chung, *PLoS One*, 2017, **12**(3), e0173585.
97. M. D. Aaslyng, L. Duedahl-Olesen, K. Jensen and L. Meinert, *Meat Sci.*, 2013, **93**, 85.
98. F. Alarcon, M. E. Baez, M. Bravo, P. Richter and E. Fuentes, *Talanta*, 2012, **100**, 439.
99. K. Schütte, H. Boeing, A. Hart, W. Heeschen, E. H. Reimerdes, D. Santare, K. Skog and A. Chiodini, *Food Chem. Toxicol.*, 2012, **50**, S724.
100. V. A. G. Londoño, L. P. Garcia, V. M. Sussel and S. Resnik, *Food Addit. Contam., Part A*, 2013, **30**(9), 1573.

CHAPTER 7

3-MCPD and Glycidyl Esters in Palm Oil

BIOW ING SIM, YU HUA WONG AND CHIN PING TAN*

Department of Food Technology, Faculty of Food Science and Technology, Universiti Putra Malaysia, 43400 Serdang, Selangor, Malaysia
*E-mail: tancp@upm.edu.my

7.1 Introduction

3-Mono-chloropropane-1,2-diol esters (3-MCPD) and glycidyl esters are now classified as a new class of food contaminants after their widespread identification in the food chain. Their occurrence in the food chain is of great concern because these fatty acid esters are believed to be completely hydrolyzed by lipases in the body into their free forms, 3-MCPD and glycidol, which are carcinogenic and can jeopardize human health.[1] Both of these compounds possess different toxicological properties. 3-MCPD has been classified as a non-genotoxic threshold carcinogen whose toxicity primarily targets the kidneys and male fertility, whereas glycidol is known to be a genotoxic, non-threshold carcinogen in animal studies.[3] Glycidol has been shown to be mutagenic in a range of *in vitro* and *in vivo* genotoxicity tests and was able to induce a broad spectrum of genotoxic effects, including gene mutations, chromosomal aberrations, sister chromatid exchange and unscheduled DNA synthesis.[2] There have been a number of animal toxicological studies examining bound MCPD, and it was revealed that it exhibited a similar degree

Food Chemistry, Function and Analysis No.19
Mitigating Contamination from Food Processing
Edited by Catherine S. Birch and Graham A. Bonwick
© The Royal Society of Chemistry 2020
Published by the Royal Society of Chemistry, www.rsc.org

of sub-chronic toxicity in rat kidneys and epididymis as that observed for 3-MCPD.[3] The European Scientific Committee on Food has suggested a tolerable daily intake (TDI) of 2 μg kg^{-1} body weight per day, and the Joint FAO/WHO Expert Committee on Food Additives has also set a provisional maximum tolerable daily intake of 2 μg kg^{-1}.[4] There are still no legal specifications but rather only trade specifications for 3-MCPD and glycidyl esters in the food market; for example, it has been suggested that refined oils for standard food applications contain 3-MCPD and glycidyl esters levels of less than 2 ppm, while the highest permitted level for refined palm oil used in infant food is less than 0.5 ppm.[5]

Palm oil is derived from the fruits of the oil palms. Due to its availability, versatility and relatively low cost, palm oil and its derivatives have become common ingredients in many consumer products. Unfortunately, refined palm oil has been found to contain relatively high average levels of 3-MCPD and glycidyl esters in comparison to other refined vegetable oils such as rapeseed, soy, and sunflower oils.[6–8]

Most commercial infant formulas are fortified with refined palm oil due to the important role of palmitic acid in infant nutrition. It has been shown that palmitic acid in the centre (sn-2) position of triacylycerol is more efficiently digested and absorbed and approximately 70% of triacylycerol molecules in human milk have palmitic acid esterified in the sn-2 position.[9] According to a pilot survey of 3-MCPD and glycidyl esters in infant formula on the market, there have been reductions in chloropropanol esters in comparison with earlier investigations in the literature, but there has not been total elimination.[10,11] Spungen and co-workers[12] suggested in their recent report that infant exposures to 3-MCPD and glycidyl esters in the USA could be further reduced if the infant formula manufacturers used non-palm-based oil or used palm-based oil with low contents of 3-MCPD and glycidyl esters. They found that the estimated mean exposures from consumption of infant formula ranged from 1–14 to 1–3 μg kg^{-1} body weight (bw) per day for 3-MCPD and glycidyl esters, respectively. The paper also indicated that the infant exposure to 3-MCPD esters in the USA are higher than those estimated for European infants, whereas the estimated infant exposures to glycidyl esters are comparable between the US and the European region. In another separate study, Arisseto *et al.*[13] reported that the average 3-MCPD intakes for infants from 0 to 5 months and 6 to 11 months are 2.49 and 1.05 μg kg^{-1} bw per day, respectively; while the average glycidyl ester intakes for infants in these two age groups are 3.65 and 1.54 μg kg^{-1} bw per day, respectively. The presence of these esters in infant products could pose a serious health issue given that many infants are solely bottle-fed with these products, which may entail considerable health risks.

In view of the increase in the global production of refined edible oils, most notably, refined palm oil, as well as their popularity among food manufacturers and households, the safety issues involving these products are garnering worldwide attention. In addition, refined palm oil is the major dietary source of both 3-MCPD and glycidyl esters because palm oil is widely used in various food products.[14] Thus, the development of mitigation strategies for 3-MCPD and glycidyl esters during refined edible palm oil production is necessary to further improve the quality and safety of this commodity to reduce any potential health risks to worldwide consumers. Until recently, research has primarily focused individually on the formation and mitigation of 3-MCPD or glycidyl esters,[15–27] while only a few studies have examined both compounds simultaneously.[28–33]

7.2 Palm Oil and Its Physicochemical Properties

Palm oil is an edible vegetable oil extracted from the mesocarps of the fruits of oil palms, primarily *Elaeis guineensis*. However, in Malaysia, *Tenera* (a hybrid of *Dura* X *Psifera*) oil palm is more commonly planted due to its relatively shorter height and good fruit bunches with higher oil content.[34]

Palm oil primarily consists of acylglycerols, of which triacylglycerols (TAG) comprise the major component (94–98%) along with small amounts of monoacylglycerols (MAG) and diacylglycerols (DAG) as metabolites resulting from the biosynthesis of TAG and as products from lipolytic activity. Like other oils, palm oil may consist of minor and variable portions of non-acylglycerol components, which can be classified as oil-soluble and oil-insoluble compounds. The oil-soluble non-acylglycerols are primarily free fatty acids (FFAs), phospholipids, tocopherols/tocotrienols, carotenoids, sterols, trace metals and some oxidation products, whereas the majority of oil-insoluble impurities are fruit fibers, nut shells and free moisture.

TAG is an ester composed of glycerol molecules bound with three fatty acids. The fatty acid chains present in palm oil can vary in their chain length (number of carbons) and structure (saturated or unsaturated), which essentially determine the chemical and physical characteristics of palm oil. As reported by Siew,[35] the chain lengths of fatty acids in palm oil range from 12–20 carbons and exist in almost equal balance between saturated and unsaturated fatty acids. The major fatty acids in palm oil are palmitic and oleic fatty acids.

The various fatty acid types that bind at different positions in the glycerol molecule can lead to a mixture of different TAGs. The different molecular TAG structures with their different chemical characteristics manifest the majority of the physical characteristics of palm oil at different temperatures, most notably the temperatures involving melting point and crystallization behavior. Solids present in the oil at any given temperature are caused by the process of crystallization that occurs in the oil; thus, palm oil can be separated under controlled thermal conditions into two components: a solid (stearin) and a liquid (olein) fraction.

The minor components of palm oil consist of partial acylglycerols (MAGs and DAGs), FFAs, phosphatides, tocopherols and tocotrienols, carotenoids and chlorophylls, sterols, fatty alcohols and vitamins. Although the minor components together constitute less than 1% of total palm oil content, they play very crucial roles in mediating the stability and refining ability as well as the nutritional value of the oil. For palm oil processers, phospholipid, carotenoid and FFA contents are of the greatest concern because if there is insufficient or improper handling throughout palm oil processing, the effects of these minor components on the quality of the final palm oil product can be significant.[36]

7.3 3-Chloro-propane-1,2-diol (3-MCPD) Esters

7.3.1 Physical and Chemical Properties

MCPDs are chlorinated analogues of glycerol with a chlorine atom in position 1 or 3. 3-MCPD esters are the esterified forms of the parent chloropropanediols, and they can be either mono- or di-esters as shown in Figure 7.1.

3-MCPD is a colorless, slightly oily liquid with a faint and pleasant odor. It has a very high boiling point of 213 °C and a density of 1.3218 g cm^{-3} (20 °C). 3-MCPD is able to undergo a wide range of chemical functionality such as alkyl halides, aliphatic carboxylic acid esters and aliphatic alcohols.[37–39] Thus, 3-MCPD readily reacts with acids, alcohols, aldehydes, ammonia, amino compounds, ketones and thiols. Degradation as well as detection of the 3-MCPD compound can be easily manipulated by understanding its chemical properties. For example, given that 3-MCPD is relatively unstable in aqueous alkaline media and is readily decomposed to glycerol, this reaction can be applied to reduce the levels of 3-MCPD in commercial hydrolyzed vegetable protein (HVP). The unstable properties of 3-MCPD in alkaline media also pose a potential problem in terms of the indirect determination of 3-MCPD ester.

7.3.2 Formation Mechanisms

There are primarily four mechanisms for the formation 3-MCPD ester, all involving S_N2 nucleophilic attack by chloride ions, as proposed by Rahn and Yaylayan.[40] The differences between these mechanisms are based on the nature of the substrate or the leaving group. Two of the proposed mechanisms involve direct nucleophilic attack by the chloride ion at two different

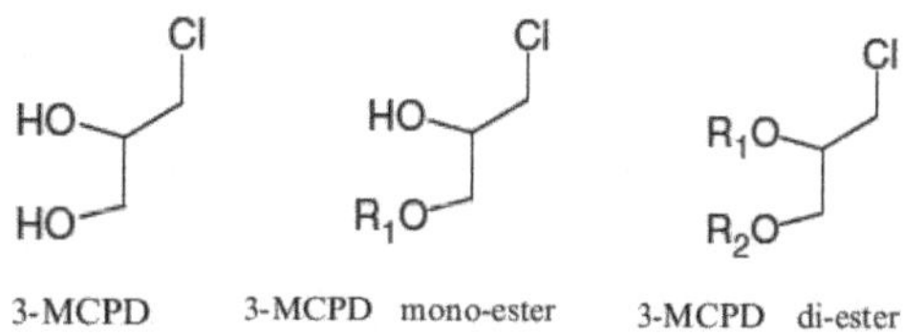

Figure 7.1 Structures of 3-MCPD, 3-MCPD mono-ester and 3-MCPD diester.

sites: either glycerol carbon atoms carrying an ester group (i) or a protonated hydroxyl group. (ii) These mechanisms are believed to occur more rapidly in acid media in the presence of partially hydrolyzed lipids. It is worth noting that under these conditions, pathway (ii) is favored over pathway (i) because water is a superior leaving group compared with fatty acids. Phosphate groups are also good leaving groups; thus, phospholipids may also be considered precursors of 3-MCPD esters. Additionally, Freudenstein *et al.*[30] reported that raw lecithin is positively correlated with the formation of 3-MCPD esters. However, refined lecithin has no significant effect on ester formation. Thus, it is speculated that the residual acids from the acid degumming process as well as the partial acylglycerols that exist in the crude lecithin are the major culprits for the formation of 3-MCPD esters and its related compounds. Furthermore, S_N2 is a reaction that involves the displacement or replacement of a leaving group (electron-deficient electrophile) by a nucleophile (consisting of a lone pair of electrons), whereby the strength of the nucleophile is greatly affected by steric hindrance. Thus, chloride ions are presumably preferred on sn-3 atoms rather than sn-2 atoms. This explains why 3-MCPD esters are present in higher abundance than 2-MCPD ester isomers.

The other two formation mechanisms are proposed to involve the formation of reactive intermediates, *i.e.* cyclic acyloxonium ions or epoxide ring, prior to nucleophilic attack by chloride ions. The cyclic acyloxonium ions can be formed through the internal nucleophilic attack of the ester carbonyls in TAG, DAG and MAG, followed by acid-catalyzed separation of the leaving group from the glycerol backbone. The leaving group can either be a protonated hydroxyl group or a carboxylic group. The neighboring ester groups act as electron donating groups that stabilize the acyloxonium ion, favoring its formation through delocalization of the positive charge. Longer fatty acid chains in a TAG increase its electron donating ability to better stabilize the resulting acyloxonium ion and displace the protonated hydroxyl group or the smaller chain fatty acid, which is a good leaving group due to its smaller size. Acyloxonium ion formation not only activates lipids towards nucleophilic attack much more effectively than esters but also partially removes the steric interference of the neighboring fatty acid chains,[40] potentially explaining the generation of 3-MCPD esters in edible oil. Conversely, 3-MCPD esters have been proposed to form through the ring opening of the epoxide ring in the form of glycidol prior to nucleophilic attack by chloride ions. Thus, glycidyl esters can be converted into 3-MCPD esters in the presence of chloride ions. This mechanism undoubtedly exists, as Sonnet[41] has reported the formation of 3-MBPD esters through the reaction of glycidyl esters with bromide ions.

Zhang *et al.*[42] proposed a novel free radical-mediated chemical mechanism for 3-MCPD ester formation at high temperature under low-moisture conditions. Using electron spin resonance (ESR) spectroscopy, the presence of free radicals in the oil at 120 °C was confirmed, whereas greater intensity of the

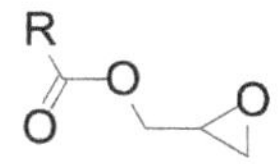

Figure 7.2 Structure of a fatty acid ester of glycidol (R = alkyl chain).

ESR signal was observed with higher temperature treatment. The involvement of the ester carbonyl groups in 3-MCPD ester formation and the reaction between sn-1,2-stearoylglycerol and chlorine gas as well as the detection of cyclic acyloxonium free radicals further supports the free radical mechanism theory.

7.4 Glycidyl Esters

7.4.1 Physical and Chemical Properties

As depicted in Figure 7.2, glycidyl ester is present only in monoester form, and the types of fatty acids bound to glycidol confer its structural diversity; in other words, glycidyl ester contains only one fatty acid moiety. Like 3-MCPD, glycidol is also a colorless liquid with a very high boiling point of 167 °C and a density of 1.143 g cm^{-3} (25 °C).[43] It possesses alkylating properties and thus is able to directly react with cellular nucleophiles due to its electrophilic epoxide structure. Glycidol has low stability (epoxide structure) and tends to react with other compounds *via* nucleophilic ring-opening and attack; thus, it is hard to determine.[44]

7.4.2 Formation Mechanisms

The proposed formation mechanism of glycidyl ester from DAG (as shown in Figure 7.3) suggests that glycidyl ester is formed predominantly by intramolecular rearrangement, followed by fatty acid elimination from DAG under high temperature. The elimination of a fatty acid from DAG begins with the abstraction of hydroxyl group protons by the adjacent carboxyl group. The generated acyloxonium intermediate is then rearranged through charge migration, resulting in the release of a fatty acid. The oxirane ring can then be formed *via* nucleophilic reaction of the alkoxide group.[45]

MAG has also been reported to be a potent precursor of glycidyl esters under high processing temperatures greater than 200 °C.[45] The authors proposed a similar mechanism to that of DAG, in which glycidyl ester is formed through an intermolecular rearrangement mechanism, followed by the elimination of a water molecule from MAG, as shown in Figure 7.4. However, due to the low levels of MAG in refined oils[46] as well as MAG volatilization during high temperature deodorization, this formation route does not appear to significantly contribute to the overall formation of glycidyl ester during edible oil refining processes.

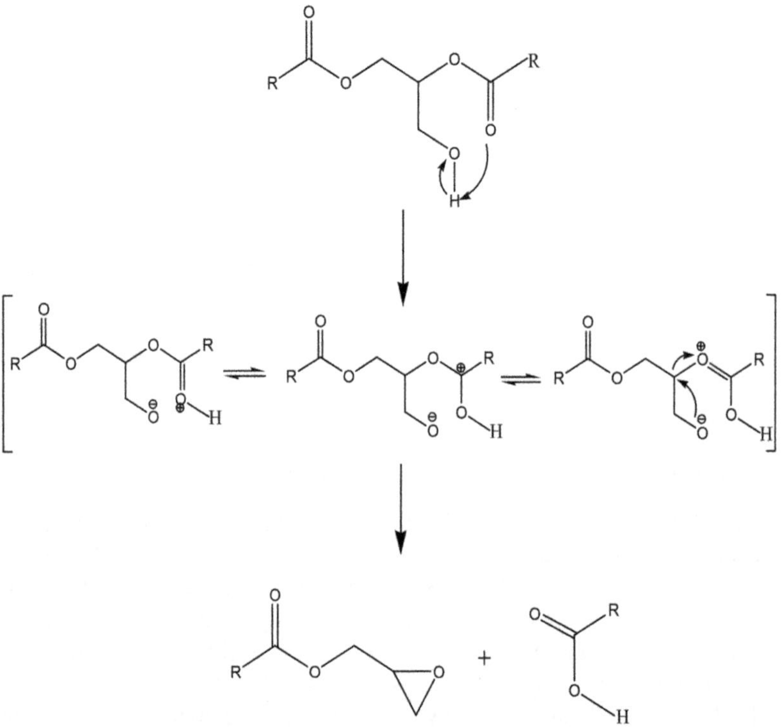

Figure 7.3 Proposed mechanism for the formation of glycidyl ester from diacylglycerol at high temperature. Reproduced from ref. 45 with permission from Elsevier, Copyright 2012.

Additionally, the formation of glycidyl ester appears to be closely related to 3-MCPD ester formation. It has been proposed that glycidyl ester may act as an obligatory precursor in the 3-MCPD ester formation pathway during food processing. Thus, glycidyl ester may form from the MAG-derived acyloxonium ion as an intermediate step in the pathway to 3-MCPD monoesters.[47] It is possible that, in the absence of chloride ions during the refining process (deodorization), the pathway may end at the glycidyl ester stage.[48] This hypothesis has been assessed by Destaillats *et al.*,[45] who heated pure MCPD mono-palmitate in hexadecane to 235 °C for 20–80 min. Analysis indicated the formation of glycidyl palmitate, whose levels gradually increased over time compared with MCPD monoesters. Glycidyl palmitate levels eventually reached an equilibrium state, and approximately 20% of MCPD monoesters were converted into glycidyl esters. However, as MCPD monoesters are not present in high abundance in refined oils,[49] it can be hypothesized that this reaction route is less significant for the formation of glycidyl esters during edible oil production.

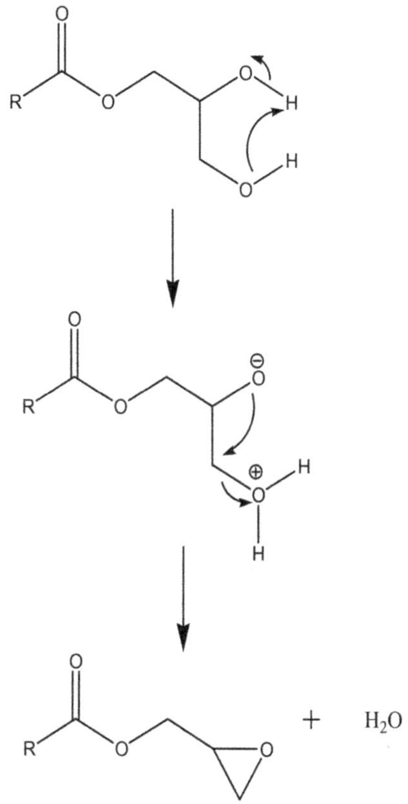

Figure 7.4 Proposed mechanism for the formation of glycidyl ester from monoacylglycerol at high temperature. Reproduced from ref. 45 with permission from Elsevier, Copyright 2012.

7.5 Overview of the Refining of Crude Palm Oil

7.5.1 Palm Oil Refining

The crude palm oil (CPO) extracted from fresh fruit bunches inevitably contains variable amounts of non-glyceride components, most of which are undesirable and must be removed. The impurities are primarily phosphatides, FFAs, trace metals, traces of pesticides, oxidation products, carotenoids and odoriferous substances such as aldehydes and ketones. The refining process is thus a crucial step in palm oil processing; it converts the crude oil into quality edible oil in the most efficient manner by removing all objectionable impurities to minimum levels while undertaking the least possible damage to TAG.

There are two major refining routes for palm oil: chemical refining and physical refining. The major difference between the two methods is the way in which FFAs are removed from the oil. Chemical refining utilizes alkali to neutralize FFAs, while physical refining involves the use of steam distillation under high temperature with vacuum conditions for FFA removal. Physical refining is most commonly used by refiners, as it has the benefits of reduced oil loss, reduced use of chemicals, reduced water and effluent handling, reduced environmental impact, higher oil yield and higher efficiency. However, to obtain good quality oils by physical refining, it is vital for oil to have a phosphorus content lower than 5 ppm prior to steam stripping. Palm oil fulfills this requirement as it contains relatively high FFA contents and low levels of phosphatides in its crude form; thus, the physical refining process has been well developed as the most feasible refining process for palm oil.[50]

7.5.2 Physical Refining

Physical refining is a continuous process comprising a two-step operation involving pre-treatment followed by steam distillation. Pretreatment involves the process of degumming CPO with concentrated phosphoric acids followed by a bleaching process involving bleaching earth. The purpose of pretreatment is to eliminate those impurities that might result in a low-quality oil product when heated to the temperatures required for steam distillation during deodorization. For example, phosphatides will break down at high temperature during deodorization, resulting in darkening and off-flavor issues.[51]

7.5.3 Pretreatment (Degumming and Bleaching)

The degumming process aims to remove particles, impurities, and the majority of phosphatides, carbohydrates, proteins and trace metals. The crude oil is treated with food-grade processing aids (*e.g.*, citric acid or phosphoric acid) and/or water at a temperature of approximately 90–100 °C, which leads to the hydration of most phosphatides such that they are insoluble in oil and readily separable. There are several types of degumming processes: water degumming, acid degumming, dry degumming, enzymatic degumming, modified acid degumming and membrane filter degumming. Among these, dry degumming is most commonly used in palm oil refining. The principal of dry degumming involves the acid decomposition of metal ion/phosphatide complexes followed by mixing with bleaching earth.[52] For CPO, 85% concentrated phosphoric acid is first added. After holding for a short reaction time (15–30 min), the CPO is then treated with 1–2% bleaching earth under a vacuum at temperatures between 80 and 100 °C. The phosphoric acid acts on non-hydratable phosphatides by precipitating them before being adsorbed by bleaching earth. In addition to adsorbing the precipitated phospholipids, bleaching earth also functions to remove any excess phosphoric acid present after degumming, such as soaps, trace metals, moisture, portions of carotenoids and other pigments and oxidation products. Sometimes, calcium carbonate is added after the bleaching earth to ensure the complete removal of residual phosphoric acid *via* neutralization. Finally, the slurry containing the oil and spent earth is removed by filtration.[53]

7.5.4 Deodorization

Deodorization is a vacuum steam distillation process that takes place at an elevated temperature at which FFAs and minute levels of volatile odoriferous components such as aldehydes and ketone are removed to obtain bland and odorless oil.[52] Additionally, deodorization also acts to decolorize the oil by thermally decomposing the residual carotenoid pigments that are present, resulting in a lighter colored oil; this is called the heat bleaching effect. During palm oil deodorization, the primary operation involves the stripping of volatile materials and thermal action from the synergistic effects of efficient vacuum, high temperature, and superheated steam. The vacuum system facilitates the distillation of FFAs and odoriferous substances at low temperature. At low absolute pressure, the boiling point of the fatty acids and the vapor pressure of the odoriferous substances will be decreased, thus resulting in a more efficient distillation process. The vacuum system is responsible for generating the required low absolute pressure and also aids in minimizing the risks of oil oxidation due to air exposure. A high deodorization temperature of 240–270 °C is necessary to increase the vapor pressure of the volatile impurities in the oil to a convenient level, resulting in shorter deodorization times. Last but not least, adequate amounts of stripping steam are required to effectively enhance the distillation of volatile components, most notably the less volatile compounds with higher boiling points. To consistently expose new oil surfaces to low absolute pressure, agitation of the oil is achieved by carefully distributing the stripping steam in the oil.[53]

7.6 Analytical Methods for 3-MCPD and Glycidyl Esters

Direct and indirect analytical methods are two general approaches used to quantify 3-MCPD and glycidyl esters. The direct methods determine individual 3-MCPD and glycidyl esters levels, while indirect methods determine free MCPD and glycidol that has been liberated from its derivative analogues.[54,55] Both methods have their benefits and drawbacks. The direct methods, as expected, have the advantage of being more straightforward because they require a minor degree of sample preparation and thereby have little chance of being impaired by errors originating from cumbersome preparation procedures.[54] In addition, direct methods provide detailed information regarding the chemical structures of the esters, which is highly useful for further toxicology research. However, a potential disadvantage is that several reference and standard compounds must be available for proper quantification.[56] To complicate things even further, unknown glycidyl derivatives, which may be present in any samples or matrix, might not be detected because there is no knowledge regarding their analytical characteristics. In contrast, indirect analytical methods permit the acquisition of all MCPD and glycidyl derivatives as long as they undergo transesterification to release their free forms. In addition, the accumulation of a whole group of compounds in one analyte should have the advantage of facilitating lower detection limits in comparison with the single mother compound.[57] Nevertheless, a major drawback of indirect detection is the need for complex sample preparation steps involving chemical reactions that can introduce additional sources of error.

Compared with the direct methods for 3-MCPD esters, the direct methods for glycidyl ester is relatively easy to handle because analytical standards are readily available in reasonable quantities. Thus, it is apparent that the development of direct analytical methods for glycidyl esters is more advanced and established than for 3-MCPD esters because glycidyl esters contain only one fatty acid moiety. In other words, there is only one single position for fatty acid esterification in glycidyl esters. Dubois *et al.*[58] drew on theoretical considerations to conclude that the concentration of glycidyl ester in most edible oils can be adequately characterized by analyzing seven glycidyl fatty acid ester species as follows: glycidyl esters of lauric, myristic, palmitic, stearic, oleic, linoleic and linolenic acids. In 2012, the American Oil Chemists' Society (AOCS) included the direct determination of glycidyl ester among its official methods, specifically Cd 28-10: Glycidyl fatty acid esters in edible oils. This method is intended for the determination of glycidyl ester in edible oils using double solid-phase extraction and liquid chromatography-mass spectrometry (LC-MS).[59]

Despite differences between the two approaches, it is obvious that indirect methods are more convenient and affordable to most users. This is evident given that AOCS validated and benchmarked three indirect methods as official methods in 2013; these methods are: (i) Cd 29a-13: 2- and 3-MCPD Fatty Acid Esters and Glycidol Fatty Acid Esters in Edible Oils and Fats by Acid Transesterification, (ii) Cd 29b-13: Determination of Bound Monochloropropanediol- (MCPD-) and Bound 2,3-epoxy-1-propanol (glycidol-) by Gas Chromatography/Mass Spectrometry (GC/MS), and (iii) Cd 29c-13: Fatty-acid-bound 3-chloropropane-1,2-diol (3-MCPD) and 2,3-epoxi-propane-1-ol (glycidol), Determination in Oils and Fats by GC/MS (Differential Measurement).[59] These methods are chosen because they are suitable for routine monitoring, perform well and give reliable results close to the known content of carefully manufactured authentic reference standards. This valuable analytical contribution by AOCS has not only simplified communications between researchers but has also improved the quality of results. However, there are also a number of newly developed indirect methods resulting from substantial efforts to investigate the optimization and further modification of analytical steps.[57]

7.6.1 Indirect Methods

The principle underlying the indirect determination of 3-MCPD ester is the conversion of a number of individual MCPD esters of fatty acids into a single compound, 3- or 2-MCPD, which can be quantified by means of transesterification. Generally, two types of indirect methods are widely employed: alkali- or acid-catalyzed transesterification. In addition to these developments and studies investigating improved indirect methods, there have been critical findings at each analytical step in the indirect methods for the analysis of 3-MCPD ester and glycidyl esters. These are outlined in the following subsections.

7.6.1.1 Use of Different Internal Standards

It is best to use esterified forms of isotopically labelled 3-MCPD and glycidol rather than their free forms because the free forms resemble the majority of native esters in oil matrices.[47,60,61] The use of 3-chloro-1,2-propandiol-dipalmitate-d5 (PP-3-MCPD-d5) gives the most accurate results, followed by 3-chloro-1,2-propandiol-1-monopalmitate-d5 (P-3-MCPD-d5) and 3-chloro-1,2-propandiol-d5 (3-MCPD-d5). Similar findings were also reported for alkaline-catalyzed methods, and differences were explained by the relatively lower stabilities of free compounds compared with their esters in alkaline media.

7.6.1.2 Extraction of Analytes from Oil Matrices

The selection of appropriate extraction solvents is of paramount importance because the solvent affects the analyte extraction efficiency. It is known that 3-MCPD ester is lipophilic, while free 3-MCPD and glycidol are hydrophilic due to their hydroxyl groups. Thus, the ideal extraction solvent should possess two properties: (i) high extraction capacity to target analytes, and (ii) good miscibility with both oil and transesterification solvents and their reactants. Thus far, the most common solvents are tetrahydrofurans (THF), diethyl ether, methyl-*tert*-butyl ether (MTBE), and ethyl acetate (EtoAc). A study examining the recovery of 3-MCPD and its esters with different extraction solvents was conducted by Ma and co-workers.[62] EtoAc and MTBE were found to yield high recovery (>70%), followed by THF, acetone, *n*-hexane, and acetonitrile (ACN).

7.6.1.3 Acidic/Alkaline Transesterification

Transesterification in the presence of methanol (methanolysis) in both alkaline and acidic media induces the conversion of TAGs and partial acylglycerols into fatty acid methyl esters (FAME) and glycerol while allowing the transformation of 3-MCPD and glycidyl esters into their free forms, which are quantifiable. Indeed, Hamlet *et al.*[63] identified an alternative method to release the free forms *via* enzymatic hydrolysis; however, the majority of methods still favor the chemical route *via* a transesterification step.

The alkaline-catalyzed transesterification approach is relatively convenient compared with the acid-catalyzed transesterification process given its short reaction time. However, a major shortcoming of this analytical method is that 3-MCPD is unstable under alkaline conditions, giving rise to glycidol. This phenomenon negatively and significantly affects the sensitivity of the method. Hrncirik *et al.*[61] and Kuhlmann[64] both reported the degradation of 3-MCPD under alkaline conditions in two independent studies. Furthermore, the specificity of the alkaline method has been questioned when sodium chloride is used as the salting out agent.[65] Glycidol present in refined oils serves as a compound leading to the undesirable formation of additional 3-MCPD *via* the bidirectional conversion reaction of glycidol into 3-MCPD in the presence of chloride ions.[66] Conversely, the acid-catalyzed transesterification method demonstrates good robustness and specificity because it gives comparable results regardless of variations in analytical conditions, such as the use of different salting out agents as described by Ermacora and Hrncirik,[60] the use of different derivatization procedures as described by Seefelder *et al.*,[49] and differential glycidol contents in oils. No degradation of 3-MCPD was observed with acid-catalyzed transesterification.[61] However, a limitation is the long period of time (16 h) required for transesterification. The Joint Research Centre of the European Commission conducted a proficiency test investigating 3-MCPD ester analysis in oils.[67] The study found that all results obtained *via* acid transesterification methods were satisfactory, while larger variations were observed for alkaline transesterification.

In addition, it is worth noting the transesterification time required for both approaches. For acidic transesterification, due to the mild conditions applied, a very long period of time is required for transesterification. Ermacora and Hrncirik[60] investigated the effects of different transesterification times, ranging from 2 h to 20 h, on 3-MCPD determination. There was no significant difference between 4 h and 20 h; thus, it was concluded that a 4 h period is sufficient to complete 3-MCPD ester cleavage with low levels of residual TAG. It is believed that the remaining TAG does not affect analysis; rather, it will be discarded together with FAME (formed during methanolysis) during the liquid–liquid extraction step that follows transesterification. Abd. Razak *et al.*[15] also reported insignificant differences in 3-MCPD levels following transesterification times between 16 and 20 h. However, the authors observed a better recovery with an 18 h transesterification period. Conversely, transesterification time exerts a significant impact in the alkaline catalyzed method due to the poor stability of 3-MCPD in alkaline medium, as it readily transforms into additional glycidol by reacting at room temperature. Thus, rather than minimizing the transesterification time, Kuhlmann[64] optimized this method by greatly reducing the reaction temperature to −25 °C and extending the reaction time to 18 h. Under these conditions, the degradation of 3-MCPD is negligible, although the reaction time is significantly prolonged to ensure a sufficient reaction rate.

7.6.1.4 Salting Out

Salting out is an important step that aids in the separation of FAMEs from the transesterified mixture, thus facilitating the liquid–liquid extraction process. The use of sodium chloride as a salting out agent in alkaline-catalyzed transesterification was found to generate additional 3-MCPD ester from glycidol.[65] However, no effects on acid-catalyzed transesterification have been observed.[61] Ermacora and Hrncirik[60] evaluated the effects of chlorides salts on acid-catalyzed transesterification by adding an aqueous solution of NaCl and tetrabutylammonium chloride individually to oils prior to acid transesterification. A linear correlation between 3-MCPD ester formation and the concentration of chloride ions was observed for both types of salts. This observation can be explained by the fact that chloride reacts with the surging amounts of partial acylglycerol and glycerol dissociated from TAG after acid-catalyzed transesterification and gives rise to 3-MCPD. As such, the authors suggest the incorporation of liquid–liquid oil extraction using demineralized water prior to transesterification; the results are promising with this method because all chlorine is completely removed, and thus 3-MCPD ester is not overestimated.

7.6.1.5 Derivatization

Derivatization is one of the most critical steps for determining 3-MCPD and glycidyl esters because it offsets the undesirable characteristics (low volatility and high polarity) of 3-MCPD and glycidyl esters, significantly increasing sensitivity through the formation of volatile derivatives. It also helps to improve specificity by permitting the monitoring of specific fragment ions from the derivatives. Phenylboronic acid (PBA) and heptafluorobutyryl imidazole(HFBI) are widely known as the two most common derivatization agents, followed by acetone, 3-pentanone or 4-heptanone.[14,68] However, all of these agents have limitations: derivatization with HFBI requires very strict anhydrous conditions because it is sensitive to moisture; excessive PBA is detrimental to the chromatographic system;[64,65] and aliphatic ketone derivatives suffer from a characteristic of fragmentation of the molecular ion from these derivatives.[69] Furthermore, this type of derivatization must be completed with the use of *p*-toluenesulfonic acid as a catalyst, which requires an additional step for its removal from the reaction medium.

Interestingly, Becalski *et al.*[70] established a new derivatization system utilizing cyclohexanone as a derivatizing agent and sulfonated polymer as a solid phase acidic catalyst to compensate for various shortcomings of the current derivatization scheme. The authors used cyclic ketone (cyclohexanone) instead of the aliphatic form to improve the quality of the spectral data, and a solid-phase catalyst (on a silica support) was used to eliminate the laborious removal steps. Moreover, the resulting derivatives (2-MCPD

1,3-dioxane and 3-MCPD 1,3-dioxolane) are stable in storage and separate well on nonpolar GC columns. The reported results for derivatization with cyclohexanone and HFBI were comparable.

During the early years of method development, all derivatization procedures with PBA were generally carried out at high temperatures of 80 °C or higher.[71] However, recent analytical methods tend to adopt lower derivatization temperatures *e.g.*, ambient temperature.[55] A high derivatization temperature was found to lead to the overestimation of 3-MCPD content deriving from undesirable glycidol formed during alkaline transesterification. Glycidol is unstable at high temperatures and is believed to undergo immediate epoxide ring opening, resulting in derivatization and, unfortunately, detection as a 3-MCPD derivative. Under ambient temperature, such unwanted side reactions can be avoided; instead, PBA effectively derivatized diols and gave comparable sensitivity and precision compared with derivatization performed at high temperature.[72] Additionally, a derivatization time of 15 min is adequate to obtain a maximum analyte signal response for derivatization at ambient temperature. This is also the case for ultrasonic-assisted derivatization, which only requires 5 min.[55] To date, several indirect methods for the determination of 3-MCPD and glycidyl esters are available as listed in Table 7.1.

Table 7.1 Indirect analytical methods for MCPD and glycidyl esters.

Method	Analyte	Comments	Reference
Acidic (H_2SO_4/MeOH)	3-MCPD esters	Long transesterification time of 16 h Good robustness High specificity	Seefelder *et al.*[49] Svejkovská *et al.*[73] Zelinková *et al.*[74]
Alkaline, mild (Société Générale de Surveillance, SGS) (NaOH/MeOH)	3- and 2-MCPD esters, glycidyl esters	Long transesterification time of 18 h Low temperature (−22 °C) applied to eliminate undesirable conversion of MCPD into glycidol Complex analytical protocol Subfreezing conditions are strictly required	Kuhlmann[64]
Acidic alcoholysis ($CH_3OH/H+$) prior to rapid alkaline catalysed trans-esterification and subsequent ester cleavage by $NaOCH_3$	3-MCPD esters and glycidyl esters	The optimized condition for transesterification is 1 min and derivatization for 15 min, both at ambient temperature. It is to minimize the undesirable conversion of glycidyl esters and 3-MCPD esters into glycidol.	Kuster *et al.*[72]

7.6.2 Direct Method

There are several direct analytical methods that have been established not only for the determination of 3-MCPD esters alone but also in combination with glycidyl esters. Direct analytical methods for glycidyl esters are more established than for 3-MCPD esters given the simple analyte pattern for glycidyl ester, which contains only one fatty acid moiety. Thus, there is only a single position for fatty acid esterification in glycidyl ester. Dubois *et al.*[58] concluded that based on theoretical considerations glycidyl ester content in most edible oils can be adequately characterized *via* the analysis of seven glycidyl ester species: glycidyl esters of lauric, myristic, palmitic, stearic, oleic, linoleic and linolenic acids. A compilation of the direct analytical methods for 3-MCPD and glycidyl esters developed in recent years is provided in Table 7.2.

7.6.3 Alternative Method

Recently, Fourier transform infrared spectroscopy (FTIR) coupled with chemometrics has been used as an alternative technique for rapid quantification of the 3-MCPD esters in palm olein.[86] The critical absorption bands used in this study are observed in the range of 700–800 cm^{-1} for C–Cl stretching vibrations. Both principal component analysis (PCA) and partial least squares (PLS) are used to analyse data obtained from the 700–800 cm^{-1} regions. The results obtained by FTIR are comparable to the results obtained using an indirect determination method by GC-MS. Frying oils with different contents of 3-MCPD esters are successfully separated into different clusters using PCA. Additionally, a model developed using PLS is able to accurately determine and predict the 3-MCPD ester levels in frying oils (Figure 7.5).

7.7 Factors Affecting 3-MCPD and Glycidyl Esters Formation in Edible Oil

Chlorine substrates are undisputedly the precursors that are involved in the formation of 3-MCPD and glycidyl esters in refined oil. There is much conjecture as to their presence throughout the oil refining process. Nagy *et al.*[87] adopted a mass defect-based filtering approach to determine the source of halogenated contaminants and their precursors in crude and partially/fully refined palm oils; they reported the presence of organic and inorganic chlorinated substances at every stage of palm oil processing and unexpectedly found them in the palm fruit itself before harvest.

The massive use of fertilizers to encourage oil palm growth and increase bunch yield has contributed to the occurrence of inorganic chlorine-containing compounds, such as potassium chloride and ammonium chloride, in palm oil.[88] Iron(III) chloride is another chlorine-containing substance that is a known coagulant used in water treatment; thus, it is introduced

Table 7.2 Direct analytical methods for MCPD and glycidyl esters.

Instruments	Analyte	Chromatographic column	Mobile-phase composition	Sample preparation	Performance characteristics	Reference
Proton nuclear magnetic resonance (^{1}H NMR) spectroscopy	Glycidyl esters	—	—	Dilution of samples in *d*-chloroform and transfer to 5 mm NMR tube → standing for 3 min to remove the bubble prior to HNMR analysis.	Recoveries: 93.20 ± 0.41% for palm olein; 100.65 ± 1.81% for DAG-enriched oil. LOD: 0.0511 mmol L LOQ: 0.170 mmol L^{-1}	Song *et al.*[75]
UPLC-Q-TOF MS (qualification); HPLC (quantification)	3-MCPD esters	UPLC-Q-TOF MS: Acquity C18 column (5 cm × 2.1 mm,1.7 mm particle, Waters)	Isolation step: Solvent A: acetonitrile Solvent B: Isopropanol	Liquid–liquid partitioning of samples in acetonitrile → centrifuge →the clear layer is evaporated and reconstituted in hexane; isolation of 3-MCPD esters on C18 column	Recoveries: 1,2-Dioleoyl-3-chloropropanediol: 94.7% ± 13.3–106.6% ± 11.8 1-Stearoyl-3-chloropropanediol: 87.5% ± 5.3–108.6% ± 3.7	Zhou *et al.*[76]
		HPLC: Hedera Si column (4.6 × 200 mm, Media: 10 nm, 3 mm particle, Hanbon Sci & Tech)	UPLC-Q-TOF MS analysis: Solvent A: Methanol Solvent B: 10 mM aqueous sodium acetate solution containing 0.1% formic acid	Using Waters 1525 HPLC system with Waters 2424 evaporative light scattering detector (ELSD).	LOD: 1,2-Dioleoyl-3-chloropropanediol: 3.43 μg mL^{-1}, 1-Stearoyl-3-chloropropanediol: 2.55 μg mL^{-1}	
			HPLC analysis: Solvent A: Hexane Solvent B: Isopropanol		LOQ: 1,2-Dioleoyl-3-chloropropanediol: 5.71 μg mL^{-1}, 1-Stearoyl-3-chloropropanediol: 5.66 μg mL^{-1}	

LC-MS/MS with electrospray ionization (ESI) in positive-ion mode	Eleven 2-MCPD diesters and three 2-MCPD monoesters	Pursuit XRs C18 column, 2.0 mm × 150 mm with 3.0 µm particles (Agilent)	For 2-MCPD monoesters: Solvent A: 2 mM ammonium formate/0.05% formic acid in 75:25 MeOH/H_2O Solvent B: 2 mM ammonium formate/0.05% formic acid in 98:2 isopropyl alcohol (IPA)/water	Dilution of samples in 20% EtOAc/MTBE; For 2-MCPD monoesters: double solid phase extraction (SPE) of samples on C18 (eluent ACN) followed by 500 mg/3 mL Si cartridge (eluent 20% EtOAc/hexane)	Recoveries and RSD: 2-MCPD diesters: 79–106% (3–13% RSD) 2-MCPD monoesters: 72–108% (4–17% RSD)	MacMahon, *et al.*[54]
			For 2-MCPD diesters: Solvent A: 2 mM ammonium formate/0.05% formic acid in 92:8 MeOH/H_2O Solvent B: 2 mM ammonium formate/0.05% formic acid in 98:2 IPA/H_2O	For 2-MCPD diesters: double SPE of samples on Si cartridge (eluent 2% Et_2O/hexane)→ SPE on C18 and Si SPE cartridge (eluent 40% EtOAc/ACN); →Evaporation of eluent and reconstitution of residue in IPA.	LOQ: 2-MCPD diesters: 30 ng g^{-1} 2-MCPD monoesters: 90 ng g^{-1}	
GC-MS	Glycidyl esters	15 m × 0.25 mm InertCap Pure WAX Carbowax column, film thickness: 0.50 µm (GL Sciences); splitless injection	—	Dilution of samples in acetonitrile → centrifuge → heptane clean-up→ back-extract the glycidyl ester (GE) from heptane phase with ACN → the solvent is evaporated to dryness, rinsed with chloroform, dried again and reconstituted in hexane/2-propanol (85:15, v/v) → isolation of GE by normal-phase liquid chromatography (NPLC)	LOD: 0.01 mg kg^{-1} per individual GE LOQ: 0.04 mg kg^{-1} per individual GE Recoveries: 85–115% depending on the level and the chain identity of the GE.	Steenbergen *et al.*[77]

(*continued*)

Table 7.2 (*continued*)

Instruments	Analyte	Chromatographic column	Mobile-phase composition	Sample preparation	Performance characteristics	Reference
LC-MS/MS with electrospray ionization (ESI)	Nine 3-MCPD monoesters and seven glycidyl esters	Pursuit XRs C18 column, 2.0 × 150 mm, with 3.0 µm particles (Agilent)	Quantification of GEs and screening for 3-MCPD monoesters: Solvent A: 2 mM ammonium formate/0.05% formic acid in 92:8 $MeOH/H_2O$ Solvent B: 2 mM ammonium formate/0.05% formic acid in 98:2 IPA/H_2O Quantification of sn-1 and sn-2 3-MCPD monoesters: Solvent A: 2 mM ammonium formate/0.05% formic acid in 75:25 $MeOH/H_2O$ Solvent B: 2 mM ammonium formate/0.05% formic acid in 98:2 IPA/water	Dilution of samples in 20% EtOAc/MTBE → double SPE of samples on C18 (eluent ACN) followed by 500 mg/3 mL Si cartridge (eluent 20% EtOAc/hexane) → evaporation of eluent and reconstitution of residue in IPA.	Recoveries and RSDs: GEs: 84–115% (3–16% RSD) sn-1 3-MCPD monoesters: 95–113% (1–10% RSD) sn-2 3-MCPD monoesters: 76.8–103% (5.1–11.2% RSD) LOD: GEs, sn-1 3-MCPD monoesters and sn-2 3-MCPD monoesters are ≤20 ng g^{-1}, 20 ng g^{-1} and 60 ng g^{-1}, respectively. LOQ: GEs, sn-1 3-MCPD monoesters and sn-2 3-MCPD monoesters are ≤30 ng g^{-1}, 60 ng g^{-1}, 180 ng g^{-1}, respectively.	MacMahon, *et al.*[7]
LC-MS/MS with APCI positive ion	Five glycidyl esters	Gemini C18 column, 250 × 2.0 mm with 5 ìm particle size (Phenomex)	100% methanol	Dilution of samples in acetone →double SPE of samples on Sep-Pak C18 cartridge (eluent MeOH) followed by Sep-Pak silica cartridge (eluent *n*-hexane/ethyl acetate 95:5, v/v) → evaporation of eluent and reconstitution of residue in MeOH/isopropyl alcohol 1:1, v/v	Recover: 84–108%	Becalski *et al.*[78]

				For analytes <0.5 mg kg^{-1}, Dilution of 0.5 g sample in 5 mL of 10% diethyl ether/hexane → pre-concentration with silica column→the dried final extract is re-dissolved in 250 µl of a mixture of methanol/isopropanol (1:1, v/v).	LOD: 70–150 µg kg^{-1} and 1–3 µg kg^{-1} for each analyte using 10 mg and 0.5 g sample of oil, respectively. LOQ: 500 µg kg^{-1} and 10 µg kg^{-1} for each analytes using 10 mg and 0.5 g sample of oil, respectively.	
LC-MS/MS	Five 3-MCPD monoesters and twenty 3-MCPD diesters	Reversed phase Luna 3 u C18(2) column (2.0 mm × 50 mm, with 3.2 µm particle size) (Phenomenex)	Solvent A: Water	Dilution of samples in mixture of MTBE and ethyl acetate (4:1);	Recovery: 78–163%	Yamazaki *et al.*[79]
			Solvent B: 0.01 mol L^{-1} ammonium acetate in methanol	For 3-MCPD monoesters, double SPE of samples on C18 (eluent MeOH) followed by 500 mg/3 mL Si cartridge [eluent ethyl acetate/hexane (20:80 v/v)]	LOD: 0.021–0.068 mg kg^{-1} and 0.014–0.071 mg kg^{-1} for 3-MCPD monoesters and diesters, respectively.	
			Solvent C: MeOH Solvent D: 2-Propanol (D)	For 3-MCPD diesters, double SPE of samples on Si cartridge [eluent ethyl acetate/hexane (5:95 v/v)] followed by 900 mg C18 cartridge [eluent ethanol, acetonitrile and methanol (5:30:65 v/v)] → evaporation of eluent and reconstitution of residue in 2-propanol.	LOQ: 0.055–0.181 mg kg^{-1} and 0.037–0.186 mg kg^{-1} for 3-MCPD monoesters and diesters, respectively.	

(*continued*)

Table 7.2 (*continued*)

Instruments	Analyte	Chromatographic column	Mobile-phase composition	Sample preparation	Performance characteristics	Reference
(i) DART-orbitrapMS, ESI with positive-ion mode (ii) U-HPLC, ESI with positive-ion mode	Nine 3-MCPD diesters and three 3-MCPD mono-esters	For (ii), Kinetex® minibore C8 (50 mm × 2.1 mm id with 1.7 µm particle size) (Phenomenex)	For (ii), Solvent A: MeOH/10 mM ammonium formate mixture (90 : 10 v/v) Solvent B: 10 mM ammonium formate in methanol	For 3-MCPD diesters, isolation by silica gel column chromatography [eluent hexane/diethylether (96 : 4 v/v)] For 3-MCPD monoesters, dilution of samples in hexane →SPE of samples in aminopropyl SPE cartridge (500 mg)→ discard the eluate consisting TAG 9eluent hexane/dichloromethane/diethylether (89 : 10 : 1 v/v/v)], followed by eluent hexane/ethylacetate (85 : 15 v/v) →evaporation of eluent and reconstitution of residue in MeOH.	Recovery: 89–120% Lowest calibration levels: 2–5 µg kg^{-1}	Moravcova *et al.*[80]
SFC/QqQ MS	Thirteen MCPD diesters	Inertsil ODS-4 (250 mm × 4.6 mm id with 5 µm particle size) (GL science)	Carbon dioxide	Dilution of samples in acetone followed by methanol.	Recovery: 62.68–115.23% LOD: <0.063 mg kg^{-1} LOQ: <0.209 mg kg^{-1}	Hori *et al.*[81]
LC/TOF-MS, ESI with positive-ion mode	Three 3-MCPD mono-esters, six 3-MCPD diesters, and five glycidyl esters	Aquity UPLC BEH C18 (50 mm × 2.1 mm id with 1.7 mm particle size) and VanGuard pre-column (Waters)	Solvent A: 0.2 mM L^{-1} sodium formate in water/MeOH (85 : 15, v/v) Solvent B: 0.2 mM L^{-1} sodium formate in MeOH/water (97.5 : 2.5, v/v)	Liquid–liquid partitioning of samples in hexane and ACN→ centrifuge; for hexane phase, SPE on Sep-Pak, Plus SI cartridges, 500 mg (Waters) (eluent chloroforms) For ACN phase, SPE on Sep-Pak Plus C18 cartridges, 500 mg (Waters) (eluent ACN and ethyl acetate) →evaporation of solvents and reconstitution of residue in ACN	Recoveries: 62.6–106.3% (2.0–11.3% RSD) and 74.8–108.8% (1.5–9.4% RSD) for glycidyl esters and 3-MCPD esters, respectively. LOD: 0.06–0.16 ng mL^{-1} and 0.02–0.86 ng mL^{-1} for glycidyl esters and 3-MCPD esters, respectively. LOQ: 0.21–0.52 ng mL^{-1} and 0.08–1.86 ng mL^{-1} for glycidyl esters and 3-MCPD esters, respectively.	Hori *et al.*[82]

LC-MS, APCI in positive-ion mode	Five glycidyl esters	ODS 150 mm × 4.6 mm id with 5 µm particle size	Solvent A: MeOH Solvent B: 2-propanol	Dilution of samples in *tert*-butyl methyl ether/ethyl acetate (4 : 1 v/v); Double SPE of samples on Sep-Pak Vac RC C18 cartridge, 500 mg (Waters) (eluent MeOH) followed by Sep-Pak Vac RC Silica cartridge, 500 mg (Waters) (eluent *n*-hexane/ethyl acetate (95 : 5 v/v)); → evaporation of eluent and reconstitution of residue in methanol/2-propanol (1 : 1 v/v).	Recovery: 94.5–109.9% (0.25–4.2% RSD) LOD: 0.025–0.033 mg g^{-1} LOQ: 0.082–0.11 mg g^{-1}	Shiro *et al.*[83]
LC-MS, APCI in positive-ion mode	Five glycidyl esters	ODS 150 mm × 4.6 mm id with 5 µm particle size	Solvent A: ACN/MeOH/water (17 : 17 : 6 v/v) Solvent B: 2-Propanol	Dilution of samples in chloroform/acetone (4 : 6 v/v); Double SPE of samples on Sep-Pak Vac RC C18 cartridge, 500 mg (Waters) (eluent ACN), followed by Sep-Pak Vac RC Silica cartridge, 500 mg (Waters) (eluent chloroform) → evaporation of eluent and reconstitution of residue in methanol/2-propanol (1 : 1 v/v).	Recovery: 103.7–109.7% (<4.0% RSD) LOD: 0.67–1.2 ng mL^{-1} LOQ: 2.2–3.8 ng mL^{-1} LOQ_{oil}: 0.22–0.38 lg/g	Masukawa, *et al.*[84]

(*continued*)

Table 7.2 (*continued*)

Instruments	Analyte	Chromatographic column	Mobile-phase composition	Sample preparation	Performance characteristics	Reference
LC-TOFMS, ESI, positive-ion mode	Five glycidyl esters and twenty 3-MCPD mono- and diesters	Luna C18 column, 50 mm × 3 mm id with 3 µm particle size (Phenomenex)	Solvent A: 0.26 mM methanol–sodium acetate solution/ MeOH/ACN (1:8:1 v/v) Solvent B: 0.26 mM methanol–sodium acetate solution/ methylene chloride/ ACN (1:8:1 v/v)	Dilution of samples in solvent B containing internal standards.	LOD: ≤1.69 mg kg^{-1} and ≤1.40 mg kg^{-1} for 3-MCPD monoesters and diesters, respectively. ≤0.29 mg kg^{-1} for glycidyl esters	Haines *et al.*[85]
LC-MS/MS	Seven glycidyl esters	Acquity UPLC HSS T3, 50 mm × 2.1 mm id with 1.8 µm particle size coupled with acquity UPLC HSS T3 Van Guard column, 5 × 2.1 mm id (Waters)	Solvent A: MeOH/water (3:1 v/v) with 0.1% formic acid Solvent B: 2-Propanol with 0.1% formic acid	Dilution of samples in cyclohexane/ethyl acetate (1:1 v/v) → GPC clean-up (glass column packed with Bio-Beads S-X3) → Glycidyl esters containing fraction is evaporated and reconstituted in acetone (for LC-MS analysis) or dichloromethane (for additional SPE clean-up for oils with high MAG and DAG content); additional SPE clean-up with 500 mg Si cartridge (eluent dichloromethane) → evaporated and reconstituted in acetone.	Recoveries: 68–111% LOQ: 50–100 µg kg^{-1}	Dubois *et al.*[58]

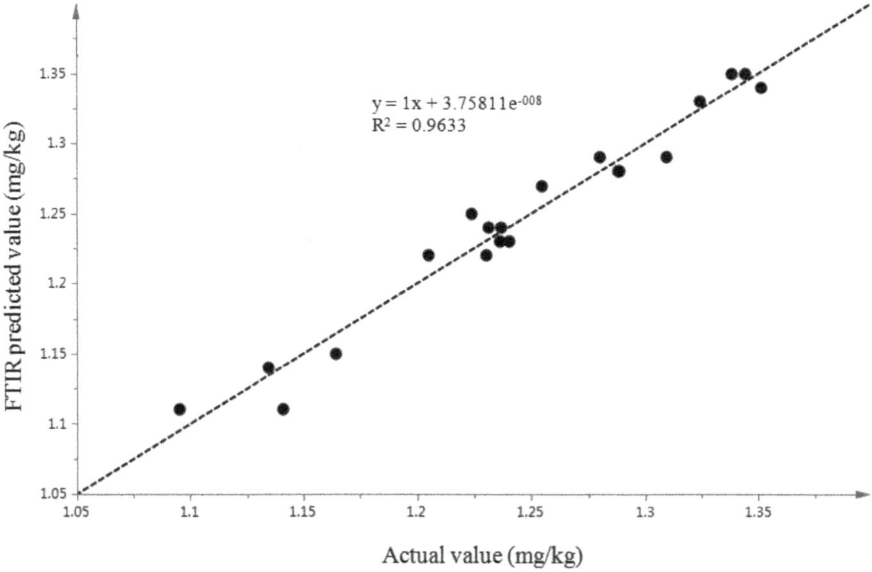

Figure 7.5 PLS calibration model for the relationship between actual value and FTIR predicted value of 3-MCPD esters in frying oils. Reproduced from ref. 86 with permission from Elsevier, Copyright 2019.

into the oil palm's environment during the irrigation process. In addition to these chlorides, calcium chloride, magnesium chloride and iron(II) chloride have also been identified.[87] Organochlorines for weed and pest control on plantations and the use of palm oil mill effluent to facilitate the recycling of sterilization condensate and water from crude oil washing increase the chances of introducing chloride-containing compounds into the palm oil production chain.[89] In addition, hydrophilic organochlorines have also been detected; these are assumed to endogenously form in oil palm fruits through the bioconversion of inorganic chlorides during the maturation stage. The hydrophilic organochlorines will then transform into more lipid-soluble forms during the sterilization of oil palm fruits. These liposoluble organochlorines will remain in the palm oil throughout subsequent palm oil refining operations. Matthäus *et al.*[19] further demonstrated that only oil-soluble chloride ions can participate in reactions with other precursors to form 3-MCPD esters. Nagy *et al.*[87] performed *in vitro* experiments to observe that these chlorinated substances were susceptible to heat and decomposed progressively upon heat treatment, simultaneously yielding increased levels of MCPD esters. Therefore, these organochlorines are believed to decompose during the high temperature deodorization step and evolve HCl, which can then react with palm oil acylglycerols, leading to the formation of MCPD esters in refined oil.[90] These data were consistent with other observations by Zhang *et al.*[42] who reported that 3-MCPD esters are formed *via* the reaction of DAG with a number of organic and inorganic chlorine sources at 120 and 240 °C through a free radical mechanism.

TAG, DAG and MAG are also possible substrates in the formation of 3-MCPD and glycidyl esters. TAGs are the major components of palm oil, and they account for approximately 88–96% of its total glycerides; DAG contents range from 4 to 12%, and MAG content is approximately 0.3%.[46] Destaillats *et al.*[91] reported that both TAG and DAG are involved in the production of 3-MCPD esters with heat treatment, and TAG can yield more 3-MCPD esters relative to DAG. Thus, the authors suggest that TAG is the primary acylglycerol class that is prone to undergoing fatty acid elimination and is involved in MCPD ester formation. Because MAG and DAG are typically minor components of TAG, they represent a workable option for the control of precursor concentrations in raw materials. However, Freudenstein *et al.*[30] studied the influence of DAG and MAG on the 3-MCPD ester contents of deodorized oil. The authors observed a positive correlation between the amount of DAG/MAG in CPO and the 3-MCPD ester content of deodorized palm oil. They also demonstrated that DAG is a stronger precursor than MAG. To investigate further, Matthäus *et al.*[19] examined the different levels of DAG in oils relative to the formation of 3-MCPD esters and related compounds. They found that 4% DAG in crude oil is a threshold value for the surging formation of 3-MCPD esters and its related components in deodorized oil. Thus, amounts of DAG exceeding 4% in crude oil will result in higher amounts of esters after heat processing, and this phenomenon is particularly apparent for glycidyl ester.[28]

Another study also reported similar findings, in which the glycidyl ester concentration may be high when DAG content is higher than 6%.[85] This finding is consistent with reports by Masukawa *et al.*,[92] who showed that commercially refined oils rich in DAG (87%) could contain glycidyl ester levels more than 10 times greater than oils with lower DAG (3.9–6.8%) contents. Because MAGs are known to be volatilized during high temperature deodorization, they are not considered potential precursors for the formation of 3-MCPD esters.[90]

In the refining of oil, a high temperature deodorization step appeared to be the most critical step for the substantial formation of 3-MCPD and glycidyl esters in refined oil, whereas the other refining steps did not seem to contribute to the ester formation.[93] Hrncirik and van Duijn[94] have revealed that 3-MCPD and glycidyl esters have very different time and temperature dependencies. The level of 3-MCPD esters in refined palm oil was found to be independent of the deodorization conditions, with regard to both temperature and residence time. On the other hand, the formation of glycidyl esters was greatly affected by the deodorization conditions, where further temperature increases above 230 °C and prolonged deodorization time led to higher levels of glycidyl esters. These findings were also in line with the results reported by Cheng *et al.*[95] On the other hand, Franke *et al.*[93] indicated that the formation of 3-MCPD esters was insignificant at a temperature of 180 °C, however, their content was doubled as the deodorization temperature increased above 240 °C.

During the degumming process, it was hypothesized that acidic conditions (from phosphoric acid) can induce the formation of 3-MCPD and glycidyl esters through the formation of acyl oxonium ions in acylglycerols.[40,96] For the bleaching step, Ramli *et al.*[20] observed a positive correlation between the acidity of the bleaching earth and the levels of 3-MCPD esters in oils, regardless of the type of degumming (acid or water degumming). In this case, acid-activated clays are believed to be the primary culprits leading to the generation of 3-MCPD esters. This generation may be caused by the acid sites, which are carried over from the residual acid-activated clay or retained after washing. These acidic clays can donate protons and induce the protonation of acylglycerol, which will then favor the formation of 3-MCPD and glycidyl esters in the presence of chloride.

7.8 Recent Mitigation Strategies of 3-MCPD and Glycidyl Esters in Palm Oil

7.8.1 Refining Processes

There have been some breakthroughs in mitigating 3-MCPD and glycidyl esters during refined palm oil production in recent years. However, the applicability of these mitigation procedures to the industry is still questionable in terms of their effectiveness for large-scale palm oil production, and there is a risk of impairing the final quality of the palm oil product.

There are three strategies that may lead to reductions in 3-MCPD and glycidyl esters, summarized in Table 7.3. The first approach is the removal of critical reactants or precursors from raw materials. This is the most promising approach to avoid alterations to refining conditions. At this juncture, there is a need to investigate the precursors that might contribute to the formation of both 3-MCPD and glycidyl esters. Chlorine is the main compound that is involved in the formation of 3-MCPD and glycidyl esters. In view of the large impact exerted by chlorine substances, various investigations have focused on the elimination of these chlorine compounds *via* water washing processes. Nagy *et al.*[87] and Craft *et al.*[90] suggested that a practice of removing chlorinated substances during the early stages of palm oil production would be more achievable and easier before the initial relatively polar and chlorinated palm oil constituents are converted into lipophilic state further down the processing chain. However, the drawbacks of water washing steps are: (i) requirement for centrifugal separation techniques in the refining plant, which is uncommon for tropical oil refining; (ii) environmental issues due to the need for high amounts of water; and (iii) oil and yield loss during washing processes. As mentioned in the previous subsection, DAG is widely known as the precursor for 3-MCPD and glycidyl esters formation and it is the by-product of the enzymatic hydrolytic degradation of TAG by TAG lipases. This process can be accelerated by harvesting, transportation and storage under elevated temperatures and harsh conditions. Thus, in this context, mitigation strategies should target the reduction of DAG and MAG concentrations in raw materials. Lipases will not be inactivated until fresh fruit bunches are sterilized, which is usually performed post-reception at the CPO mill. Therefore, it is best to reduce

Table 7.3 Summary of possible mitigation strategies for 3-MCPD and glycidyl esters in refined oil.

Strategies	Methods	Findings	Reference
Removal of precursors from raw materials	Washing of CPO with ethanol : water (1 : 1)	~30% 3-MCPD ester reduction	Craft *et al.*[90]
	Washing of palm fruit pulp before oil extraction	~95% 3-MCPD ester reduction	Matthäus *et al.*[19]
Optimization of the refining process	*Degumming*		
	Incorporation of water degumming toglycidyl esterther with acid degumming	~80% 3-MCPD ester reduction	Zulkurnain *et al.*[27]
	Neutralisation of phosphoric acid in degummed oil with:	~45% 3-MCPD ester reduction	Matthäus *et al.*[89]
	Potassium hydroxide		
	Sodium hydroxide	~35% 3-MCPD ester reduction	
	Calcium oxide		
	Bleaching		
	Use of natural or acidic bleaching clays with neutral pH	~35% glycidyl ester reduction	Matthäus *et al.*[19]
	Deodorization		
	Generating strip steam by formic acid		
	Use of refining aids:		
	Glycerol		Craft *et al.*[90]
	Ethanol		Matthäus *et al.*[89]
	Diacetin	~25% 3-MCPD ester reduction	
	Carbonates	~35% 3-MCPD ester reduction	
	Dual deodorization process:	~50% } MCPD esters and related compounds reduction ~66% } MCPD esters and related compounds reduction	
	Short term deodorization (5 min) at Higher temperature (270 °C) followed by long-term deodorization (120 min) at lower temperature (200 °C).	80% 3-MCPD ester and glycidyl ester reduction	
	Short-path distillation		Matthäus and Pudel;[89] Pudel *et al.*[31]
		90% 3-MCPD ester and glycidyl ester reduction	

the time period between harvesting and palm fruit milling as much as possible. In addition, Hamlet *et al.*[96] also suggested breeding new species of oleiferous plants that possess lower lipase activities, but the applicability and implementation of this approach are highly dependent on technological development and public acceptability.

The second approach is to change or optimize the refining process, for example by modifying the refining parameters and/or adding a processing step or refining aids. It is best to avoid this option because it is difficult to compromise between the optimization of refining processes for 3-MCPD ester or glycidyl ester formation and the properties of the final oil product. An attempt to incorporate water degumming and acid degumming was made to eliminate precursors of 3-MCPD and glycidyl esters and sufficiently remove both gums and impurities simultaneously.[27] It is known that water degumming alone is inadequate for the complete removal of non-hydratable phosphatides in palm oil because these substances are oil-soluble and have a very low affinity for water. Thus, acid degumming is a vital processing step that cannot be omitted because non-hydratable phosphatides can only be removed by acids (phosphoric acid or citric acid) that are stronger than phosphatidic acid; otherwise, the non-hydratable phosphatides will darken the oil, and off-flavor problems will develop at the deodorization step. Zulkurnain *et al.*[27] also discovered that the water degumming step exerted the most significant effect on the reduction of 3-MCPD esters: higher water dosages resulted in larger effect size. Comparatively, deodorization temperature and acid dosage did not play a substantial role in this process. This observation can be explained by the fact that the water degumming process permits precursors to be washed off at the start, lessening subsequent chances of 3-MCPD ester formation during later steps. In addition, a reduced degumming temperature and acid dosage can also result in a considerable decline, approximately 70%, in 3-MCPD ester levels in oil.[27] A significant decrease in the 3-MCPD ester levels of bleached oil was observed when a lower dose of phosphoric acid was used for degumming.[20] However, increased acid-degumming temperature and acid dosage can improve oil oxidative stability by removing impurities such as bound metal ions that are associated with non-hydratable phospholipids; they can also assist in degrading carotene compounds, which can then contribute to further color removal. In addition, Ramli *et al.*[20] discovered that neutralizing the phosphoric acid in degummed oil with calcium oxide or calcium carbonate prior to bleaching reduced 3-MCPD ester levels in refined oil. After neutralization, the neutral oil is then washed in hot water to remove the undesirable soapstock that has formed. This additional washing step was found to aid in the removal of undesirable impurities and precursors from the oil.

Acid-activated clays used during bleaching steps are believed to be the primary culprits leading to the generation of 3-MCPD esters. Thus, Ramli *et al.*[20] recommended the use of natural bleaching clays or acid-activated clays with a more neutral pH. The primary difference between these two bleaching earths is that natural bleaching earth in aqueous suspension is slightly acidic or neutral, whereas activated bleaching earth is highly acidic.[97] Neutral bleaching clays are aluminum silicates that contain relatively high quantities of calcium, magnesium or iron and are usually activated by heat treatment. Because the high metal content restricts the adsorptive activity of these clays, acid treatment is used to remove the metals from reactive sites, permitting the so-called acid-activated clays to possess a much higher adsorptive capacity.[52]

Because high temperature deodorization is the major culprit underlying 3-MCPD and glycidyl esters formation, minimizing the temperature will exert some impact in reducing these contaminants. Pudel *et al.*[98] and Craft and Nagy[29] suggested a deodorization temperature below 240 °C, which is expected to impart beneficial effects in terms of contaminant reduction, most notably glycidyl ester. However, it is widely known that high temperature steam distillation removes FFA and minute levels of volatile odoriferous components such as aldehydes and ketones, inactivates color compounds (carotenoids), and decomposes pesticide residues. Thus, lower deodorization temperatures must be validated to ensure that all quality parameters for the refined product are met. In contrast, Matthäus *et al.*[19] recommended the use of formic acid to generate stripping steam during deodorization, rather than water, to reduce glycidyl ester. This notion was based on the hypothesis that the glycidyl ester is very unstable due to its reactive epoxide binding structure and tends to degrade under acidic solutions. Furthermore, a stepwise increase in the mitigation efficiency was observed with increasing formic acid concentrations. However, 3-MCPD ester is virtually unaffected by acid treatment. Alternatively, Craft and Nagy[29] suggested the addition of refining aids such as ethanol or glycerol to refined bleached palm oil because it was anticipated that these low-molecular-weight alcohols would lead to the formation of volatile chlorinated adducts that would be stripped from the oil during deodorization and subsequently result in a decrease in 3-MCPD ester formation.

A third approach is the removal of 3-MCPD and glycidyl esters after processing; however, the impact of the removal process on the quality of the resulting oil must be taken into account. This option may not be preferred by oil processors because it involves an additional cost and will decrease overall profit. Bornscheuer and Hesseler[99] were the first to suggest an enzymatic approach to remove 3-MCPD esters from an aqueous and biphasic system (in the presence of oil) by converting 3-MCPD into non-toxic glycerol, which is permitted in food. The basic hypothesis is that the fatty acid monoester of 3-MCPD is hydrolyzed by lipase to release the free form before conversion into the corresponding epoxide glycidol by halohydrin dehalogenase (HHD). This product is then ultimately hydrolyzed by an epoxide hydrolase (EH) into glycerol.

This approach enables the easy exchange of the food contaminant 3-MCPD for non-toxic glycerol; however, the technical application of this method in palm oil plants has yet to be assessed and approved. Strijowski *et al.*[100] investigated the removal of 3-MCPD ester in refined palm oil using adsorbent material. Synthetic magnesium silicate and calcinated zeolite were found to be capable of reducing the contents of these contaminants and their related substances, the majority of which were identified as glycidyl ester but not 3-MCPD ester. Synthetic magnesium silicate required higher temperatures (80 °C and above) and treatment times of 30 min or more to achieve an apparent reduction in 3-MCPD esters and related substances, whereas calcinated zeolite reacts relatively quickly and consistently over a wide range of treatment temperatures and times. The advantages of these approaches are that they exert no adverse effects on the palm oil in terms of sensory properties and oxidative stability, treatment with calcinated zeolite imparts trivial improvements in oil quality (stability), and they can easily be implemented within a short time frame.

Sim *et al.*[33] reported that acid-activated bleaching earth has a significant effect on the reduction of 3-MCPD esters in refined, bleached and deodorized (RBD) palm oil as compared with natural bleaching earth or acid-activated bleaching earth with neutral pH. This finding indicated that the performance and adsorption capacity (*e.g.* total pore area, $m^2\ g^{-1}$) of the bleaching earth are more important than the acidity of the bleaching earth in the removal of 3-MCPD esters. In addition, the same study also reported that a combination high dose of phosphoric acid during degumming and subsequent treatment with acid-activated bleaching earth can eliminate almost all glycidyl esters from palm oil. The authors postulated that the elimination of glycidyl esters is due to structural modification of glycidyl esters under acidic conditions.

Short-path distillation has been used as an alternative refining process for crude palm oil for the mitigation of 3-MCPD and glycidyl esters.[31] Although no 3-MCPD esters and low glycidyl esters are obtained after short-path distillation, the sensory quality of palm oil is significantly impaired. The authors suggested mild deodorization of the treated palm oil in order to improve its sensory quality.

7.8.2 Fat Modification and Storage

Recently, Kyselka *et al.*[101] evaluated the mitigation of 3-MCPD and glycidyl esters in RBD palm oil *via* wet fractionation and hydrogenation processes. This study concluded that fractionated palm stearin had significantly lower contents of 3-MCPD esters (71–74%) and glycidyl esters (21–26%) than fractionated palm olein. While the hydrogenation process triggered rapid degradation of 3-MCPD esters (51–70%) in palm oil products. The type of nickel catalyst has an important role in the reduction of glycidyl esters (1–42%) during the hydrogenation process. Figure 7.6 shows a typical degradation of 3-MCPD esters using Raney nickel or nickel catalysts.

Storage of oil can be an option to reduce glycidyl esters in palm oil.[102] Storage of oil samples at various temperatures ranging from −20 °C to 15 °C has been shown to degrade the glycidyl esters in RBD palm oil. RBD palm oil stored at refrigerated temperature (5 °C) shows the greatest reduction of

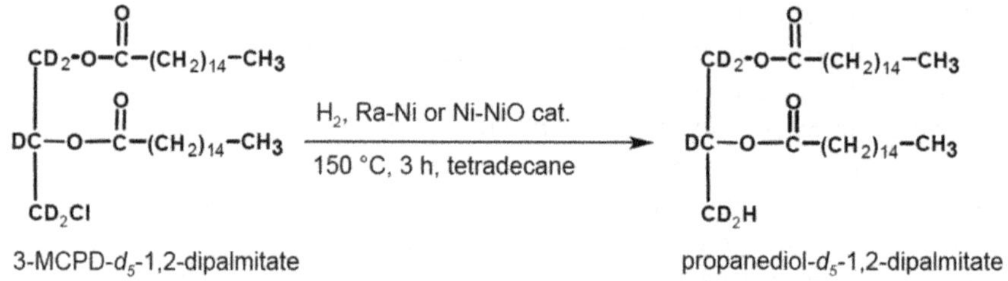

Figure 7.6 Degradation of 3-MCPD-d5-1,2-dipalmitate using Raney nickel or nickel catalysts. Reproduced from ref. 101 with permission from Springer Nature, Copyright 2018.

glycidyl esters after 340 days of storage. The authors concluded that the crystallization process of RBD palm oil is the main reason for this phenomenon. Crystallization of oil is an exothermic process and this energy is sufficient to cleave the epoxy structure of glycidyl esters. However, in the same study, the storage of RBD palm oil shows no significant effect on the degradation of 3-MCPD esters.

7.8.3 Deep-fat Frying

Deep-fat frying is a popular food preparation technique and palm olein is frequently used as a heating medium for this purpose. Many researchers have evaluated the effect of deep-fat frying on formation of 3-MCPD and glycidyl esters in palm oil products. Recently, Ben Hammouda *et al.*[103] observed no endogenous formation of 3-MCPD and glycidyl esters in blended oil samples (olive pomace oil and refined palm oil) after 16 h of frying of French fries. Others have made the same observation regarding the formation of 3-MCPD and glycidyl esters during deep-fat frying of potato chips using high-oleic sunflower oils.[104] Similar results have been reported by Yıldırım and Yorulmaz.[105] They evaluated the effect of rosemary extract and NaCl on deep-fat frying of potato chips using sunflower oil. Lower contents of 3-MCPD glycidyl esters with addition of rosemary extract in sunflower oil are reported after prolonged frying.

Recently, a systematic study aimed at investigation of the formation of 3-MCPD and glycidyl esters in palm olein during the deep-fat frying of high carbohydrate and high protein foods has been carried out by Wong and coworkers.[106,107] In addition, the mitigation of these process contaminants by the addition of a synthetic or natural antioxidant to frying oils and the development of an alternative technique for the estimation of 3-MCPD ester levels in palm olein by using FTIR coupled with chemometric analysis are also reported by the same research group.[86,108]

The frying duration, frying temperature and concentration of sodium chloride have been shown to have significant effect on the formation of 3-MCPD and glycidyl esters in RBD palm olein during the deep fat frying of potato chips.[106] The levels of 3-MCPD and glycidyl esters in palm olein with repeated frying are shown in Figures 7.7 and 7.8. For 3-MCPD esters, the trend is a decrease when the frying duration increases and an increase when the frying temperature or concentration of NaCl decreases. For glycidyl esters, the trend is increasing when the frying temperature, frying duration or concentration of NaCl increased.

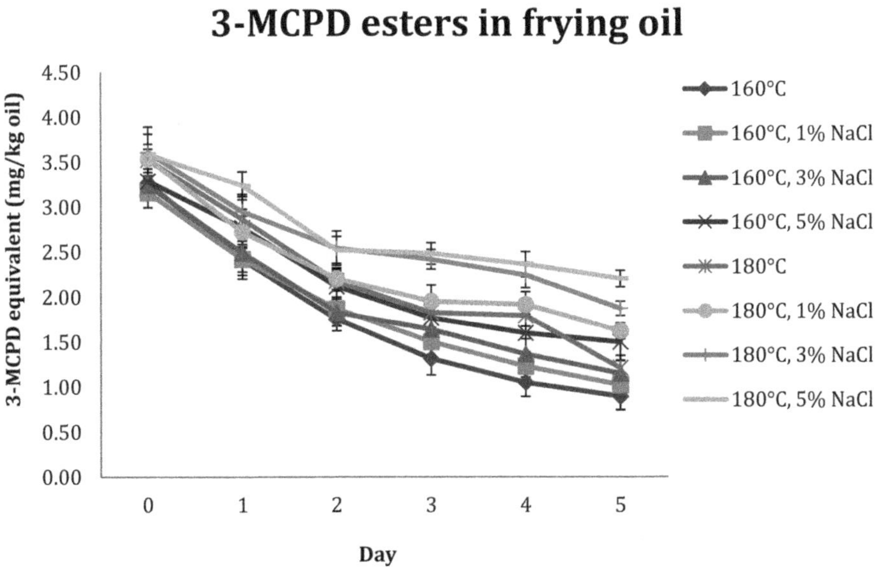

Figure 7.7 3-MCPD esters of frying oil in different frying systems (potato chips). Reproduced from ref. 106 with permission from Elsevier, Copyright 2017.

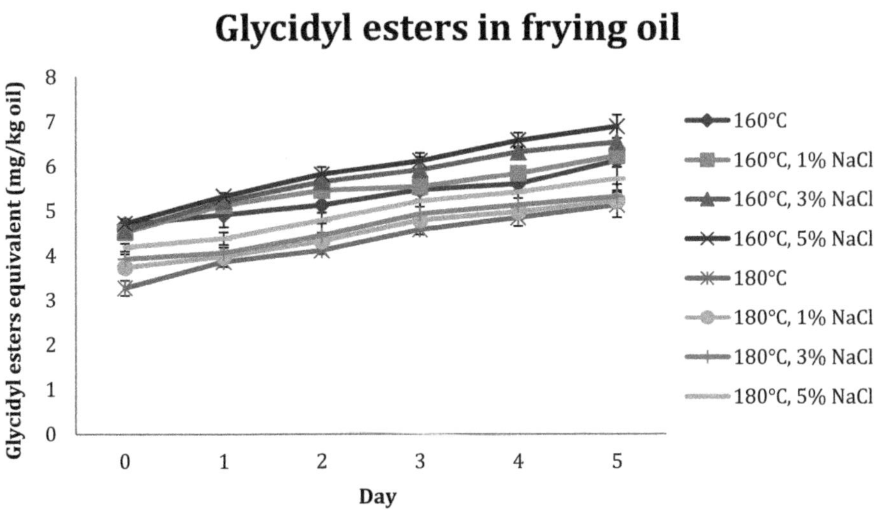

Figure 7.8 Glycidyl esters of frying oil in different frying systems (potato chips). Reproduced from ref. 106 with permission from Elsevier, Copyright 2017.

In another study with a similar experimental setup, chicken breast meat was used as a high-protein food.[107] The 3-MCPD esters decomposed (Figure 7.9) while the glycidyl esters (Figure 7.10) were relatively stable and increase slightly at high temperature over 5 days of frying. Losses of 5–15%

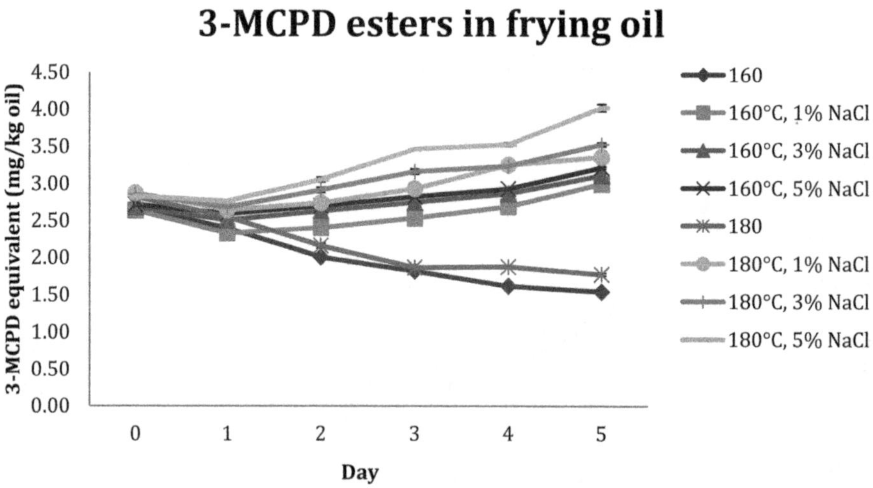

Figure 7.9 3-MCPD esters of frying oils in different frying systems (chicken breast meat). Reproduced from ref. 107 with permission from John Wiley and Sons, Copyright © 2017 American Oil Chemists' Society (AOCS).

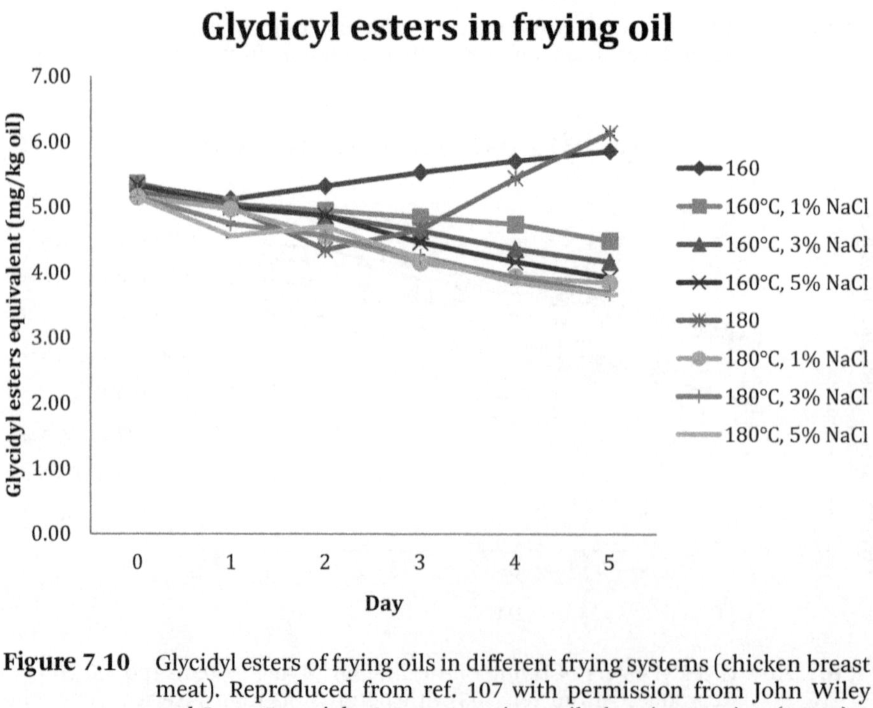

Figure 7.10 Glycidyl esters of frying oils in different frying systems (chicken breast meat). Reproduced from ref. 107 with permission from John Wiley and Sons, Copyright © 2017 American Oil Chemists' Society (AOCS).

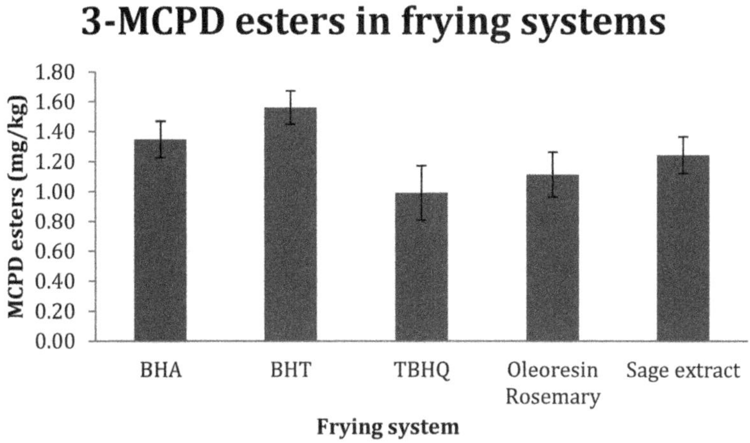

Figure 7.11 The concentration of 3-MCPD esters (mg kg^{-1}) in frying oils with the added of BHA, BHT, TBHQ, oleoresin rosemary and sage extract, respectively. Reproduced from ref. 86 with permission from Elsevier, Copyright 2019.

of 3-MCPD were reported in this study over the entire frying duration at two selected frying temperatures (160 and 180 °C). 3-MCPD esters are not stable and decompose under long heating duration. The authors suggested that the degradation rate of glycidyl esters is lower than the formation rate during the frying processes. However, the 3-MCPD esters increased in the presence of sodium chloride in frying oils at elevated temperature, while the glycidyl ester decreased slightly. After 5 days of frying, the oil quality was within the safety limit but showed slightly higher degradation levels than the frying oil of fried potato chips.

The addition of synthetic [butylated hydroxyanisole (BHA), butylated hydroxytoluene (BHT) and tert-butyl hydroquinone (TBHQ)] or natural (oleoresin rosemary, sage extract) antioxidants have been shown to improve the oil stability and lower the levels of 3-MCPD and glycidyl esters in RBD palm olein.[86,108] The ability of the antioxidants to reduce the formation of 3-MCPD esters varied. Figure 7.11 shows that the order of activity found for the antioxidants is: TBHQ ≈ oleoresin rosemary > sage extract ≈ BHA ≈ BHT for frying oil.[86] TBHQ and oleoresin rosemary exhibit comparable and the best anti-oxidative effect to effectively reduce the levels of 3-MCPD and glycidyl esters.

7.9 Conclusion

For the palm oil industry, many studies have indicated that 3-MCPD and glycidyl ester formation can be reduced *via* modification of the physical refining process. In order to study the root cause of this issue and further eliminate the presence of these processed contaminants, it is worth venturing into the upstream processes such as the optimization of the palm oil milling process for possible mitigation strategies. A number of studies from our group also revealed that certain minor components of palm oil such as organochlorine substances, DAG and FFA are possible precursors for ester formation during the refining process. However, little is known about the sources of the chlorine atom and how it is introduced into the oil palm fruit. Therefore, future work should focus on this area, especially on its identification as well as it origin, so that this substance can be avoided or eliminated during further milling or refining processes. Through these comprehensive studies, a multifaceted approach can be envisaged, beginning with a reduction of the chloride application in fertilizer or pesticides, continuing with the production of crude palm oil with reduced precursors, and ending with feasible modification to the physical refining steps to obtain refined palm oil or palm oil products with extremely low levels of 3-MCPD and glycidyl esters.

References

1. International Agency for Research on Cancer, Some chemicals present in industrial and consumer products, food and drinking-water, *IARC Monogr. Eval. Carcinog. Risks Hum.*, 2012, **101**, 149–284.
2. N. Bakhiya, K. Abraham, R. Gürtler, K. E. Appel and A. Lampen, *Mol. Nutr. Food Res.*, 2011, **55**, 509.
3. S. Onami, Y.-M. Cho, T. Toyoda, Y. Mizuta, M. Yoshida, A. Nishikawa and K. Ogawa, *Arch. Toxicol.*, 2014, **88**, 871.
4. JECFA, Joint FAO/WHO Expert Committe on Food Additives, Evaluation of certain food additives and contaminants, *WHO Technical Report Series, No. 909*, 2002.
5. W. D. Greyt, Edible Oil Processing: Deodorization, *AOCS Lipid Library*, http://lipidlibrary.aocs.org/content.cfm?ItemNumber=40326, last accessed December 2018.
6. C. Li, S. P. Nie, Y. Q. Zhou and M. Y. Xie, *Food Chem. Toxicol.*, 2015, **75**, 8.
7. S. MacMahon, T. H. Begley and G. W. Diachenko, *Food Addit. Contam., Part A*, 2013, **30**, 2081.
8. R. Weißhaar, *Eur. J. Lipid Sci. Technol.*, 2011, **113**, 304.
9. U. Bracco, *Am. J. Clin. Nutr.*, 1994, **60**(6 suppl), 1002S.
10. A. Becalski, T. Zhao, S. Feng and B. P. Y. Lau, *J. Food Compos. Anal.*, 2015, **44**, 111.
11. F. Wöhrlin, H. Fry, M. Lahrssen-Wiederholt and A. Preiß-Weigert, *Food Addit. Contam., Part A*, 2015, **32**, 1810.
12. J. H. Spungen, S. MacMahon, J. Leigh, B. Flannery, G. Kim, S. Chirtel and D. Smegal, *Food Addit. Contam., Part A*, 2018, **35**, 1085.
13. A. P. Arisseto, W. C. Silva, G. R. Scaranelo and E. Vicente, *Food Control*, 2017, **77**, 76.
14. C. Crews, A. Chiodini, M. Granvogl, C. Hamlet, K. Hrnčiřík, J. Kuhlmann, A. Lampen, G. Scholz, R. Weisshaar, T. Wenzl, P. R. Jasti and W. Seefelder, *Food Addit. Contam., Part A*, 2012, **30**, 1.

15. R. A. Abd. Razak, A. Kuntom, W. L. Siew, N. A. Ibrahim, M. Ramli, R. Hussein and K. Nesaretnam, *Food Control*, 2012, **25**, 355.
16. B. D. Craft, K. Nagy, W. Seefelder, M. Dubois and F. Destaillats, *Food Chem.*, 2012, **132**, 73.
17. A. Ermacora and K. Hrncirik, *Food Chem.*, 2014, **161**, 383.
18. C. Li, Y. Zhou, J. Zhu, S. Wang, S. Nie and M. Xie, *LWT--Food Sci. Technol.*, 2016, **69**, 586.
19. B. Matthäus, F. Pudel, P. Fehling, K. Vosmann and A. Freudenstein, *Eur. J. Lipid Sci. Technol.*, 2011, **113**, 380.
20. M. R. Ramli, W. L. Siew, N. A. Ibrahim, R. Hussein, A. Kuntom, R. Abd. Razak and K. Nesaretnam, *J. Am. Oil Chem. Soc.*, 2011, **88**, 1839.
21. M. R. Ramli, W. L. Siew, N. A. Ibrahim, A. Kuntom and R. A. Abd. Razak, *Food Addit. Contam., Part A*, 2015, **32**, 817.
22. K. A. Sampaio, A. P. Arisseto, J. V. Ayala, C. Stevens, R. Ceriani, R. Verhé and A. J. A. Meirelles, *Toxicol. Lett.*, 2013, **221**(suppl.), S122.
23. M. Shimizu, J. Moriwaki, D. Shiiba, H. Nohara, N. Kudo and Y. Katsuragi, *J. Oleo Sci.*, 2012, **61**, 23.
24. M. Shimizu, K. Vosmann and B. Matthäus, *Eur. J. Lipid Sci. Technol.*, 2012, **114**, 1268.
25. J. Šmidrkal, M. Tesařová, I. Hrádková, M. Berčíková, A. Adamčíková and V. Filip, *Food Chem.*, 2016, **211**, 124.
26. M. Zulkurnain, O. M. Lai, R. A. Latip, I. A. Nehdi, T. C. Ling and C. P. Tan, *Food Chem.*, 2012, **135**, 799.
27. M. Zulkurnain, O. M. Lai, S. C. Tan, R. Abdul Latip and C. P. Tan, *J. Agric. Food Chem.*, 2013, **61**, 3341.
28. B. D. Craft, A. Chiodini, J. Garst and M. Granvogl, *Food Addit. Contam., Part A*, 2012, **30**, 46.
29. B. D. Craft and K. Nagy, *Lipid Technol.*, 2012, **24**, 155.
30. A. Freudenstein, J. Weking and B. Matthäus, *Eur. J. Lipid Sci. Technol.*, 2013, **115**, 286.
31. F. Pudel, P. Benecke, K. Vosmann and B. Matthäus, *Eur. J. Lipid Sci. Technol.*, 2016, **118**, 396.
32. M. Shimizu, P. Weitkamp, K. Vosmann and B. Matthäus, *Eur. J. Lipid Sci. Technol.*, 2013, **115**, 735.
33. B. I. Sim, H. Muhamad, O. M. Lai, F. Abas, C. B. Yeoh, I. A. Nehdi, Y. P. Khor and C. P. Tan, *J. Oleo Sci.*, 2018, **67**, 397.
34. N. A. Morad, The physical properties of palm oil mixtures for design of process equipment, University of Leeds, 2006.
35. W. L. Siew, Analysis of palm and palm kernel oils, in *Advances In Oil Palm Research*, ed. Y. Basiron,B. S. Jalani and K. W. Chan, Malaysian Palm Oil Board, Kuala Lumpur, Malaysia, 2000, pp. 968–1035.
36. M. Maclellan, *J. Am. Oil Chem. Soc.*, 1983, **60**, 368.
37. *Hawley's Condensed Chemical Dictionary*, ed. R. J. Lewis Sr., Van Nostrand Rheinhold Co., New York, 12th edn, 1993, p. 267.
38. *The Merck Index – An Encyclopedia of Chemicals, Drugs, and Biologicals*, ed. S. Budavari, Merck and Co., Inc., Whitehouse Station, New Jersey, 1996, p. 356.

39. IARC Monographs Working Group on the Evaluation of Carcinogenic Risks to Humans, 3-monochloro-1,2-propanediol (updated 14 September 2016), Some chemicals present in industrial and consumer products, food and drinking-water, 2011, p. 101.
40. A. K. K. Rahn and V. A. Yaylayan, *Eur. J. Lipid Sci. Technol.*, 2011, **113**, 323.
41. P. E. Sonnet, *Chem. Phys. Lipids*, 1991, **58**, 35.
42. X. Zhang, B. Gao, F. Qin, H. Shi, Y. Jiang, X. Xu and L. Yu, *J. Agric. Food Chem.*, 2014, **61**, 2548.
43. *Handbook of Data on Organic Compounds*, ed. D. R. Lide and G. W. A. Milne, CRC Press, Inc., Boca Raton, 3rd edn, 1994, vol. I, p. 3801.
44. Dixie Chemical Co., *Product Technical Bulletin: 2,3-Epoxy-1-propanol (Glycidol)*, Houston, 1995.
45. F. Destaillats, B. D. Craft, M. Dubois and K. Nagy, *Food Chem.*, 2012, **131**, 1391.
46. E. M. Goh and R. E. Timms, *J. Am. Oil Chem. Soc.*, 1985, **62**, 730.
47. R. Weißhaar and R. Perz, *Eur. J. Lipid Sci. Technol.*, 2010, **112**, 158.
48. M. Habermeyer, S. Guth and G. Eisenbrand, *Eur. J. Lipid Sci. Technol.*, 2011, **113**, 314.
49. W. Seefelder, N. Varga, A. Studer, G. Williamson, F. P. Scanlan and R. H. Stadler, *Food Addit. Contam., Part A*, 2008, **25**, 391.
50. MPOC, The Next Big Thing: Pakistan's Physial Oil Refining Industry, *Malaysia Palm Oil Fortune*, 2008, vol. 6, p. 10..
51. Y. Basiron, *Palm Oil in Bailey's Industrial Oil and Fat Products*, John Wiley & Sons, Inc., New York, 2005, pp. 231–261.
52. V. Gibon, W. De Greyt and M. Kellens, *Eur. J. Lipid Sci. Technol.*, 2007, **109**, 315.
53. R. D. O'Brien, *Fats and Oils: Formulating and Processing for Applications*, Taylor & Francis, Bota Raton, 3rd edn, 2010, pp. 73–191.
54. S. MacMahon, C. D. Ridge and T. H. Begley, *J. Agric. Food Chem.*, 2014, **62**, 11647.
55. A. Ermacora and K. Hrncirik, *J. Am. Oil Chem. Soc.*, 2013, **90**, 1.
56. K. Miyazaki, K. Koyama, H. Sasako and T. Hirao, *J. Am. Oil Chem. Soc.*, 2012, **89**, 1403.
57. Z. Zelinkova, A. Giri and T. Wenzl, *Food Control*, 2017, **77**, 65.
58. M. Dubois, A. Tarres, T. Goldmann, G. Loeffelmann, A. Donaubauer and W. Seefelder, *J. Agric. Food Chem.*, 2011, **59**, 12291.
59. AOCS, *Official Methods and Recommended Practices of the AOCS, Cd 29a-13, Cd 29b-13 and Cd 29c-13*, AOCS, Urbana, 7th edn, 2017.
60. A. Ermacora and K. Hrncirik, *J. Am. Oil Chem. Soc.*, 2012, **89**, 211.
61. K. Hrncirik, Z. Zelinkova and A. Ermacora, *J. Lipid Sci. Technol.*, 2011, **113**, 361.
62. F. Ma, P. Li, B. Matthäus, W. Zhang and Q. Zhang, *J. Sep. Sci.*, 2012, **35**, 2241.
63. C. G. Hamlet, P. A. Sadd and D. A. Gray, *J. Agric. Food Chem.*, 2004, **52**, 2059.
64. J. Kuhlmann, *Eur. J. Lipid Sci. Technol.*, 2011, **113**, 335.
65. R. Weißhaar, *Eur. J. Lipid Sci. Technol.*, 2008, **110**, 183.

66. N. Kaze, H. Sato, H. Yamamoto and Y. Watanabe, *Eur. J. Lipid Sci. Technol.*, 2011, **113**, 1168.
67. L. Karasek, T. Wenzl and F. Ulberth, *Eur. J. Lipid Sci. Technol.*, 2011, **113**, 1433.
68. C. G. Hamlet and P. A. Sadd, Chloropropanols and Chloroesters, *Process-induced Food Toxicants*, John Wiley & Sons, Inc., New York, 2008, pp. 175–214.
69. F. M. Dayrit and M. R. Niñonuevo, *Food Addit. Contam.*, 2004, **21**, 204.
70. A. Becalski, T. Zhao and D. Sit, *Food Energy Secur.*, 2013, **2**, 157.
71. DGF, DGF Standard Method C III 18 (09) – Determination of ester-bound 3-chloropropane-1,2-diol (3-MCPD esters) and 3-MCPD forming substances in fats and oils by means of GC-MS, 2009.
72. M. Küsters, U. Bimber, S. Reeser, R. Gallitzendörfer and M. Gerhartz, *J. Agric. Food Chem.*, 2011, **59**, 6263.
73. B. Svejkovská, O. Novotný, V. Divinová, Z. Réblová, M. Doležal and J. Velíšek, *Czech J. Food Sci.*, 2004, **22**, 190.
74. Z. Zelinková, B. Svejkovská, J. Velíšek and M. Doležal, *Food Addit. Contam.*, 2006, **23**, 1290.
75. Z. Song, Y. Wang, G. Li, W. Tan and S. Yao, *Eur. J. Lipid Sci. Technol.*, 2015, **117**, 918.
76. H. Zhou, Q. Jin, X. Wang and X. Xu, *Food Control*, 2014, **36**, 111.
77. H. Steenbergen, K. Hrnčiřík, A. Ermacora, S. de Koning and H.-G. Janssen, *J. Chromatogr. A*, 2013, **1313**, 202.
78. A. Becalski, S. Y. Feng, B. P. Y. Lau and T. Zhao, *Anal. Bioanal. Chem.*, 2012, **403**, 2933.
79. K. Yamazaki, M. Ogiso, S. Isagawa, T. Urushiyama, T. Ukena and N. Kibune, *Food Addit. Contam., Part A*, 2012, **30**, 52.
80. E. Moravcova, L. Vaclavik, O. Lacina, V. Hrbek, K. Riddellova and J. Hajslova, *Anal. Bioanal. Chem.*, 2012, **402**, 2871.
81. K. Hori, A. Matsubara, T. Uchikata, K. Tsumura, E. Fukusaki and T. Bamba, *J. Chromatogr. A*, 2012, **1250**, 99.
82. K. Hori, N. Koriyama, H. Omori, M. Kuriyama, T. Arishima and K. Tsumura, *LWT--Food Sci. Technol.*, 2012, **48**, 204.
83. H. Shiro, N. Kondo, N. Kibune and Y. Masukawa, *Eur. J. Lipid Sci. Technol.*, 2011, **88**, 356.
84. Y. Masukawa, H. Shiro, N. Kondo and N. Kudo, *J. Am. Oil Chem. Soc.*, 2011, **88**, 15.
85. T. Haines, K. Adlaf, R. Pierceall, I. Lee, P. Venkitasubramanian and M. Collison, *J. Am. Oil Chem. Soc.*, 2011, **88**, 1.
86. Y. H. Wong, K. M. Goh, F. Abas, M. Maulidiani, K. L. Nyam, I. A. Nehdi, H. M. Sbihi, M. M. Gewik and C. P. Tan, *LWT--Food Sci. Technol.*, 2019, **100**, 404.
87. K. Nagy, L. Sandoz, B. D. Craft and F. Destaillats, *Food Addit. Contam., Part A*, 2011, **28**, 1492.
88. S. Ng, Nutrition and nutrient management of the oil palm. New thrust for the future perspective, ed. N. S. Pasricha and S. K. Bansal, *Potassium for Sustainable Crop Production, International Symposium on Role*

of Potassium in India, Potash Research Institute of India, and International Potash Institute, New Delhi, 2002, pp. 415–429.
89. B. Matthäus and F. Pudel, *Lipid Technol.*, 2013, **25**, 151.
90. B. D. Craft, K. Nagy, L. Sandoz and F. Destaillats, *Food Addit. Contam., Part A*, 2011, **29**, 354.
91. F. Destaillats, B. D. Craft, L. Sandoz and K. Nagy, *Food Addit. Contam., Part A*, 2011, **29**, 29.
92. Y. Masukawa, H. Shiro, S. Nakamura, N. Kondo, N. Jin, N. Suzuki, N. Ooi and N. Kudo, *J. Oleo Sci.*, 2010, **59**, 81.
93. K. Franke, U. Strijowski, G. Fleck and F. Pudel, *LWT--Food Sci. Technol.*, 2009, **42**, 1751.
94. K. Hrncirik and G. van Duijn, *Eur. J. Lipid Sci. Technol.*, 2011, **113**, 374.
95. W. Cheng, G. Liu, L. Wang and Z. Liu, *Compr. Rev. Food Sci. Food Saf.*, 2017, **16**, 263.
96. C. G. Hamlet, L. Asuncion, J. Velíšek, M. Doležal, Z. Zelinková and C. Crews, *Eur. J. Lipid Sci. Technol.*, 2011, **113**, 279.
97. W. Zschau, *Eur. J. Lipid Sci. Technol.*, 2011, **103**, 505.
98. F. Pudel, P. Benecke, P. Fehling, A. Freudenstein, B. Matthäus and A. Schwaf, *Eur. J. Lipid Sci. Technol.*, 2011, **113**, 368.
99. U. T. Bornscheuer and M. Hesseler, *Eur. J. Lipid Sci. Technol.*, 2010, **112**, 552.
100. U. Strijowski, V. Heinz and K. Franke, *Eur. J. Lipid Sci. Technol.*, 2011, **113**, 387–392.
101. J. Kyselka, K. Matějková, J. Šmidrkal, M. Berčíková, E. Pešek, B. Bělková, V. Ilko, M. Doležal and V. Filip, *Eur. Food Res. Technol.*, 2018, **244**, 1887.
102. B. Matthäus, K. Vosmann, P. Weitkamp, D. Grundmann and H. J. Kersting, *Eur. J. Lipid Sci. Technol.*, 2016, **118**, 418.
103. I. Ben Hammouda, A. Zribi, A. Ben Mansour, B. Matthäus and M. Bouaziz, *Eur. Food Res. Technol.*, 2017, **243**, 1219.
104. A. Dingel and R. Matissek, *Eur. Food Res. Technol.*, 2015, **241**, 719.
105. A. Yıldırım and A. Yorulmaz, *Grasas Aceites*, 2018, **69**, e273.
106. Y. H. Wong, H. Muhamad, F. Abas, O. M. Lai, K. L. Nyam and C. P. Tan, *Food Chem.*, 2017, **219**, 126.
107. Y. H. Wong, O. M. Lai, F. Abas, K. L. Nyam, I. A. Nehdi, H. Muhamad and C. P. Tan, *J. Am. Oil Chem. Soc.*, 2017, **94**, 759.
108. Y. H. Wong, K. M. Goh, K. L. Nyam, I. A. Nehdi, H. M. Sbihi and C. P. Tan, *Food Control*, 2019, **96**, 488.

CHAPTER 8

Advanced Glycation Endproducts (AGEs) in Food: Health Implications and Mitigation Strategies

CATHERINE S. BIRCH* AND GRAHAM A. BONWICK

AgriFoodX Limited, Chester, UK
*E-mail: cath.birch@agrifoodx.com

8.1 Introduction

Modern diets principally undergo some form of heat-processing and as a result contain high levels of advanced glycation endproducts (AGEs). Dietary advanced glycation endproducts (dAGEs) are known to contribute to increased oxidative stress (OS) and inflammation, which are linked to the increasing incidence of diabetes, kidney disease and cardiovascular disease (CVD).[1]

Biological amines react with reducing sugars to form a complex family of rearranged and dehydrated covalent adducts that are often yellow-brown and/or fluorescent and include many cross-linked structures.[2] Food scientists have studied this process as a source of flavour, colour and texture changes in cooked, processed and stored foods.[3] For many food

Food Chemistry, Function and Analysis No.19
Mitigating Contamination from Food Processing
Edited by Catherine S. Birch and Graham A. Bonwick
© The Royal Society of Chemistry 2020
Published by the Royal Society of Chemistry, www.rsc.org

formulation scientists these attributes are desirable. During the last 30 years it has been realised that this process, known as the Maillard reaction or advanced glycation, also occurs slowly *in vivo*.[4] Long-lived proteins such as structural collagen and lens crystallins particularly are implicated as pathogenic targets of AGE processes.[5–8] Collagen fibrils are the smallest discernible structural elements of load-bearing tissues and as such, they are of high biomechanical importance. The low turnover of collagen cause AGEs to accumulate within the collagen fibrils as a part of normal ageing as well as in various pathologies.[5]

AGE formation in vascular wall collagen appears to be an especially harmful event, causing crosslinking of collagen molecules to each other and to circulating proteins. This is implicated in plaque formation, basement membrane thickening and loss of elasticity in the vascular system.[9,10] The chemistry of these later-stage, glycation-derived protein crosslinks is still not fully elucidated but is based on the hypothesis that AGE formation involves reactive carbonyl groups.

Food is heated by various cooking methods to increase palatability and reduce bacteriological risk, however, during the cooking process toxic compounds can form within the food. The wide range of toxic compounds can present a risk to human health.[11] These toxic compounds can form during food processing such as through heating, roasting, baking and grilling. Frying food is by far the most pertinent process that acts as a generator of a wide variety of toxic compounds in food. Multiple deleterious compounds are generated through frying such as, mutagenic polycyclic aromatic hydrocarbons (PAHs), acrylamide, heterocyclic aromatic amines (HAAs) and AGEs. These process derived toxic compounds are formed by the interaction of fats, proteins and natural sugars, as oxidation products are chemically bound to proteins and sugars in the food product. Food processing-derived chemical contaminants in cooked and highly processed foods are of increasing concern to consumers. Thermal processing leads to the formation of heat-induced toxic compounds that can exert adverse toxicological effects that pose a threat to human health.[12] Final concentrations of thermally-induced contaminants in food depends on the concentration of the corresponding precursors, as well as the degree and duration of thermal processing. This process is influenced by cooking time, temperature and method of cooking. Heat treatment is the most widely used method for food processing in the food industry and the home. Some 80–90% of food for consumption is processed to some degree[13] Additionally, high cooking temperatures can lead to the formation of process-related toxic contaminants in foods that are rich in fat, protein and sugars.

Food products containing lipids, especially unsaturated fatty acids, are prone to oxidation reactions, which can lead to the formation of off-flavors, due to side-reaction products that potentially have toxic effects and reduce the shelf-life of products. The Maillard reaction products (MRPs) are widely produced in foods containing reducing sugars and amino-bearing compounds during thermal processing and storage. MRPs possess excellent antioxidant ability in many food products, through chelation of metal ions, breakdown of free radical chains and hydrogen peroxide and scavenging of reactive oxygen species.[14] Protein glycation can influence foam quality as well as gelation or emulsifying properties in food products. Partial hydrolysis of proteins accompanied by anti-solvent triggered nanoparticulation before or after conjugation can enhance emulsifying properties of food products.[15] However, modification of amino acids and proteins also limit their bioavailability. Mostly, glycation takes place on free lysine or arginine residues, however, cysteine, tryptophan and histidine residues are also favourable glycation targets as lysine and tryptophan are essential amino acids. This may lower the nutritive value of a food product, so strategies to remove these compounds may increase the nutritional value of food products.

AGEs, also known as glycotoxins, are a diverse group of highly oxidant compounds with pathogenic significance in diabetes, CVD, kidney disease and in several other chronic disease states.[16] The formation of AGEs is a part of normal metabolism, but if excessively high levels of AGEs are reached in tissues and the circulatory system, they have the potential to become pathogenic.[17] The pathological effects of AGEs are related to their ability to promote OS and inflammation by binding with cell surface receptors or cross-linking with proteins, altering their structure and function.[18–22] Among the well-studied AGEs are the stable and relatively inert N^{ε}-carboxymethyllysine (CML),[22] N^{ε}-1-carboxyethyllysine (CEL) and the highly reactive derivatives of methylglyoxal (MG) that include N^{δ}-(5-hydro-5-methyl-4-imidazolon-2-yl)-ornithine (MG-H1).[23,24] MG-derived stress has recently been implicated in the development and progression of cancer through alteration of proteins and DNA.[23] Individually these AGEs can be derived from protein and lipid glycoxidation in the diet.

High-temperature heating methods promote new dAGE formation by approximately 10- to 100-fold above the uncooked stage, across a wide range of food categories.[1] Animal-derived foods that are high in fat and protein are generally AGE-rich and prone to new AGE formation during cooking. In contrast, carbohydrate-rich foods such as vegetables, fruits, whole grains and milk contain relatively few AGEs, even after cooking.

Preparation of common foods under various cooking conditions using water and heat processing conditions has a different effect on dAGE content. For example, scrambled eggs prepared over medium-low heat have about half the dAGEs of eggs prepared in the same way but over high heat. Poached or steamed chicken have less than one quarter of the dAGEs of roasted or barbequed chicken. In all food categories so far examined, exposure to high temperatures and lower moisture levels coincided with higher dAGE levels for equal weight of food when compared to foods prepared at lower temperatures or with more moisture. Furthermore, frying, barbequing, grilling and roasting yielded more dAGEs compared with boiling, poaching, stewing, and steaming, in other words preparation of food without using direct-heat approaches forms fewer AGE-related compounds. The study by Miglio *et al.*, indicated that microwaving did not raise dAGE content to the same extent as other dry heat cooking methods for the relatively short cooking times (6 min or less) that were tested.[25]

8.2 Synthesis and Formation of Advanced Glycation Endproducts

AGEs form by the interaction of aldoses with proteins and the subsequent molecular rearrangements of the covalently linked sugars, leading to a diverse group of fluorescent compounds of yellow-brown colour.[26] This heterogeneous class of non-enzymatically glycated proteins or lipids is found in the blood plasma and accumulates in blood vessel walls and tissues principally during the ageing process, potentially leading to a range of metabolic disease states.

The first, reversible step takes place between the carbonyl group of a reducing sugar and an amino terminal group of a protein, lipid or nucleic acid generating a "Schiff base".[27,28] Subsequently through structural irreversible re-arrangements, more stable keto-amines are formed, known as "Amadori products" (APs).[6,29–32] The APs undergo further physical changes through oxidation, dehydration and degradation to finally yield highly stable AGE compounds.[29,33] The carbonyl groups necessary for the reaction do not exclusively originate from carbohydrate metabolism but can also be formed during lipid or protein degradation. When the carbonyl group originates from lipid catabolism, for example malondialdehyde, the products of the reaction are termed "advanced lipoxidation endproducts" (ALEs).[34–37] However, the differentiation between AGEs and ALEs is not always possible: CML, for example, can be formed during either carbohydrate or lipid catabolism and can therefore be attributed both terms AGEs or ALEs.[33]

The APs and associated Heyns products (HPs) are formed during the early stage of the process. The subsequent degradation of the APs and HPs is dependent on the pH of the system. At neutral or acid pH, APs and HPs undergo 1,2-enolization with the formation of furfural and hydroxymethylfurfural (HMF). At alkaline pH the degradation of APs and HPs involves 2,3-enolization and formation of reductones and fission products, mainly α-hydroxycarbonyls and α-dicarbonyls.[29] Dicarbonyls react with amino acids with the formation of aldehydes and α-aminoketones, known as the Strecker degradation pathway or they incorporate nitrogen into the reaction products.[38,39] During the advanced stages of the reaction several reactions occur, including cyclization, dehydration, retroaldolization, molecular rearrangement, isomerization and further condensation. In the final stage, melanoidins and AGEs are formed.[40,41]

Reactive 1,2-dicarbonyl compounds (DCs) can be generated from carbohydrates during food processing. Recent studies have focused on chemical formation pathways and the role that DCs in food products play in physiological systems as they have been found to be mutagenic.[42–44] DCs are formed by dehydration and redox reactions and can have a significant impact on the palatability, aroma and colour of food products. The precursors of AGE formation, these compounds have been identified as 3-deoxyglucosone, 3-deoxygalactosone and glucosone that predominate over MG, glyoxal and

3,4-dideoxyglucosone-3-ene.[44–47] The absorption bioavailability of DCs is heavily influenced by the release of these components from the food matrix during the digestion process, which is in turn dependent on their reactivity with other components of intestinal fluids. However, it is evident that further research is needed to determine DCs role during intestinal digestion and potential pathophysiological effect at quantities found in real food matrices.

Moreover, the Maillard reaction initiates reactions which cause formation of highly reactive aldehydes (*e.g.* Strecker aldehydes and dicarbonyls) which further react with biomolecules to generate products such as CEL, pyrraline or MG-derived hydroimidiazolone such as MG-H1.[17]

Hyperglycaemia greatly enhances AGE formation and deposition is exacerbated in diabetes, in which their presence has been linked to secondary complications, especially microvascular disease.[48] The mechanism of AGE formation has detrimental effects and indicates a central role for cellular interactions of AGEs and the involvement of a novel receptor for AGE (RAGE).[49,50] RAGE, an immunoglobulin superfamily member, which mediates the binding of AGEs to endothelial cells (ECs) and mononuclear phagocytes, interacts with a lactoferrin-like polypeptide that also binds AGEs and appears to activate intracellular signal transduction mechanisms consequent to its interaction with the glycated ligand.[51] RAGE is expressed by ECs, mononuclear phagocytes, smooth muscle cells, mesangial cells and neurons, indicating a potential role in the regulation of their properties in homeostasis and/or their dysfunction in the development of diabetic complications and neurological conditions such as Alzheimer's disease and Parkinson's disease.[52–56] Since AGEs have been shown to generate reactive oxygen intermediates,[57] tethering of AGEs to the cell surface by their receptors focuses OS on cellular targets. This mechanism results in changes in gene expression and the cellular phenotype.[58] RAGE is an important cell surface receptor and is considered a pattern recognition receptor.[59] Additionally, RAGE is currently seen as an attractive therapeutic target with the aim of reducing these system damaging processes.

The discovery of RAGE and development of natural plant-based compounds[60] to block their interaction with AGEs should provide insights into the role of this ligand–receptor interaction in the pathogenesis of diabetic complications and other disease conditions.

8.3 Methods of Detection of Advance Glycation Endproducts

These toxic compounds are present in many different foods at relatively low amounts, often in the parts-per-billion (ppb) range and consequently analytical methods for detection are usually based on mass spectrometry with either a gas or liquid chromatographic step prior to detection. Methods for antiglycation compound measurement by *in vivo* techniques involve using two model systems: Bovine serum albumin–glucose (BSA–glucose) and bovine serum albumin–methylglyoxal (BSA–MGO).[61] About 100 compounds that have been identified can be detected as reaction products between xylose and glycine using high-performance liquid chromatography (HPLC) and thin-layer chromatography (TLC) on solvent-soluble material.[29,62,63]

Analytical methods for AGE detection have been established and AGEs were first analysed in a broad range of food products in response to their discovery. The first database with foods and their CML content was published in 2004. This was based on the measurement of CML in an ELISA format using a monoclonal antibody against CML.[64] The development of a food database demonstrated that foods with high CML content were also foods with a high protein and fat content. The initial database of food was expanded in 2010 by Uribarri and colleagues to include 549 food items.[1] This extended database also included food measurements based on cooking methods. In 2012, Hull *et al.* published a list of CML levels in foods using UHPLC-MS/MS techniques.[65] The latest comprehensive database containing CML, CEL and MG-H1 levels in foods was published by Scheijen *et al.* in 2016.[26,66] Interrogation of the latest edition of the database indicates that canned meats, heavily processed meat products or nuts and grain products processed at high temperatures had the highest levels of CML, CEL and MG-H1, in comparison to butter, coffee, fruits and vegetables which had the lowest overall AGE-derivative content.

Furthermore, there are quite wide discrepancies among selected analytical methods. For example, results from an antibody-based ELISA are not comparable with mass spectrometry techniques.[67] As an illustration of this discrepancy, CML content of high-fat products such as butter, mayonnaise and olive oil were overestimated by mass spectrometry methods. Whereas, CML levels for carbohydrate-rich foods measured by ELISA were underestimated. More recently 20 food items were analysed by HPLC-ITMS/MS and ELISA, processed meat products showed the highest levels of CML.[68] One drawback in measurement techniques is that very similar food items can significantly differ in their nutritional profile, this can lead to problems of comparability using the databases, as the wide variations in food processing techniques make it particularly difficult to accurately measure AGE levels.

From the current methods employed for AGE-related derivatives, it is obvious that a quantitative method is preferable, such as HPLC-MS/MS. Additionally the current freely available databases are a good resource for calculating the levels of AGE-related compounds. Two such databases can be found at:

https://lemchem.file3.wcms.tu-dresden.de
http://www.adajournal.org

Over recent years, aptamers have been developed for the rapid detection of AGE-related toxic derivatives.[59,69–72] Aptamers are developed as an alternative to more familiar antibody immunoassay detection methods. Aptamers are short, single-stranded DNA or RNA (ssDNA or ssRNA) molecules, that can selectively bind to a specific target, including proteins, carbohydrates, peptides, toxins and live cells.[73]

Electrochemical aptasensors for contaminant detection including AGEs have recently been developed. Aptamer-based sensors are promising because they permit multiplexed analysis that provides fast, sensitive specificity and low cost.[74] Aptamers are developed by the immobilisation of appropriate oligonucleotide sequences onto a platform electrode surface often using nanomaterials such as graphene, carbon nanotubes or metals.[75] Advanced functional materials such as aptasensors, can improve sensitivity and selectivity. Detection limits have been obtained at the nanomole (nM) to femtomole (fM) level and in general are below the maximum level accepted by European legislation for food products.[76]

In 2017 Matsui *et al.* developed an aptamer against RAGE *in vitro* to examine the effect on the development and progression of diabetic nephropathy in a rat model.[77] This successful RAGE aptamer bound to RAGE with a K_d of 5.68 nMol L^{-1} and effectively blocked the binding of AGEs to RAGE. The aptamer was further developed as a treatment strategy. Their results indicate that a continuous infusion of the developed RAGE aptamer over 6 weeks could diminish the development and progression of diabetic nephropathy and has the potential to be extended to human disease conditions.

Olsen *et al.* (2019), demonstrated a microfluidic SELEX approach to develop and isolate glycan targets.[78] Glycan-specific proteins (lectins) often exhibit low affinity and can be non-specific, these features often prove difficult to raise antibodies against. The developed aptamer rapidly indicated a high affinity of K_d <100 nMol for the glycan target. This development has the potential to generate accessible sensors for the emerging glycomics field of research.

However, there remain gaps in analytical techniques, especially the availability of rapid, affordable and reliable alternative methods with readily available standards and validated methods.[79] It is clear that difficulties relating to current methods for the detection and measurement of AGEs are of particular concern, including the lack of a universally established method of detection or a defined unit of measurement.

8.4 Advanced Glycation Endproducts and Their Involvement in Human Disease

Inflammation is considered an initial first step in a wide range of pathological conditions and is critical for the development of many complex diseases and disorders including autoimmune diseases,[80] metabolic syndrome,[81] neurodegenerative diseases,[82,83] cancers,[84–87] diabetes[88–92] and cardiovascular disease.[93–96] The inflammatory response can be considered as a two-stage process: chronic inflammation, which can be defined as a dysregulated form of inflammation[97,98] and acute inflammation,[99–101] which is defined as a regulated form. dAGEs that bind to proteins exert an inflammatory response and directly stimulate the immune system.[17,102] Establishing methods to control both acute and chronic inflammation are important for developing therapeutic treatments against various diseases and disorders.

Food preparation techniques can undergo a long chain of processing, including heating, roasting, grilling, baking, canning and fermentation, where each technique is a potential source of chemicals that can be added to food. Depending upon the food preparation process, chemicals that are mixed as additives or chemicals that are formed accidentally during food processing can have negative impacts on human health leading to a wide range of pathological conditions.

8.5 Advanced Glycation Endproducts as Endocrine Disruptors

The mammalian endocrine system is comprised of interdependent tissues that are mediated through hormones that are essential for metabolism, growth and development.[103] Humans secrete over fifty different hormones to orchestrate major physiological functions. However, these vital functions can be disrupted by a number of internal and external chemical stressors that are identified as endocrine disruptors (EDs). AGEs formed through non-enzymatic glycation derived from the diet or through metabolic changes can significantly disrupt the delicate balance of the endocrine system.[104] Although most of the EDs are man-made, some of the EDs are naturally formed in the human diet.

Highly processed convenience foods are increasingly purchased by consumers due to their taste, convenience and general low cost. The manufacture of these day-to-day foods often involves extreme temperature processing, resulting in the formation of AGEs that can independently promote OS, aging, diabetes, cancer, degenerative diseases and hormonal disruption. Furthermore, exposure to these endocrine disruptors can be caused by migration of phthalates from plastic packaging materials.[105–107] Chemical additives to foods are extremely heterogeneous, omnipresent and have a wide application in the manufacture of food products. EDs or endocrine disrupting chemicals are defined as exogenous chemicals that are capable of intervening with naturally occurring endogenous hormones with their production, release, transport, metabolism, binding, action, or elimination[104,108,109] Reports from different scientific models and epidemiological studies have implicated EDs as a significant concern to public health.[108,110,111]

Thermal processing is widely applied in the domestic as well as industrial setting to inactivate microbes and improve sensory qualities such as flavour, texture, taste and shelf-life. However, from a biochemical perspective, all forms of thermal processing promote a series of chemical and physical changes to food proteins. AGE generation is speculated to interfere with hormones such as insulin, leptin, adiponectin, parathyroid hormone, testosterone, androgen, oestrogen and progesterone.[104,112,113]

8.6 Therapeutic Strategies to Reduce Advanced Glycation Endproducts

Some 14 900 research papers have been published between 2016 and 2019 on treatment strategies for AGE related conditions (Scopus; PubMed and Web of Science data 2019). This substantial body of work indicates both the need to find either pharmacological or naturally derived compounds that counteract food-derived AGEs or those formed within the body as a consequence of disease or the ageing process. Furthermore, many natural plant compounds are of ethnopharmacological relevance and have been shown to have a beneficial health effect, most often used in what is termed 'traditional medicine'. Multiple classes of drugs are capable of quenching AGE precursors, preventing the formation of AGE-derived toxicants.[114–116] The prototypical molecules with these capabilities are arginine analogues such as aminoguanidine and pyridoxamine, which along with arginine itself, can effectively scavenge AGE precursors or Amadori product-derived AGE precursors.[117–119] Other anti-AGE drugs work by targeting the structure of glycation target proteins, making them less prone to glycation (*e.g.* limonene).[120] Pre-clinical studies with AGE inhibitors have been largely positive, with reports of amelioration of diabetic nephropathy, retinopathy, neuropathy and CVD phenotypes in multiple animal models.[6]

Studies by many research groups have shown the effectiveness of aminoguanidine as a standard therapeutic agent in slowing or preventing a wide range of complications of diabetes and aging in animals and recently, in humans.[61,121] Liu *et al.* (2018), have developed a new class of agents, exemplified by 4,5-dimethyl-3-phenacylthiazolium chloride (DPTC), that can chemically break already-formed AGE protein–protein crosslinks.[122] These pharmacological agents are based on a new theory of AGE crosslinking that postulates that alpha-dicarbonyl structures are present in AGE protein–protein crosslinks.

Pyridoxamine is a vitamin B_6 metabolite that has been exploited widely as an AGE inhibitor.[31,118] Its mode of action appears similar to aminoguanidine, in that it can quench reactive dicarbonyl and can to some extent reverse Amadori intermediates. Pyridoxamine has been tested clinically for its ability in mechanistic targeting of AGEs to diminish diabetic complications, either alone or in combination with other vitamins or vitamin analogues.[6] Currently, thc pharmacological drug of choice for the reduction and inhibition of AGEs is aminoguanidine (CH_6N_4); (Trade name: Pimagedine®). Other drugs used in the treatment of AGEs include pyridoxamine, benfotiamine, angiotensin converting enzyme inhibitors, angiotensin receptor blockers, statins, ALT-711 (alagebrium) and thiazolidinediones.[123–125] The most recent promising anti-AGE agents are statins and thiazolidinediones. These drugs treat the associated conditions of diabetic microvascular retinopathy, neuropathy and nephropathy.[126–130] Other changes to dietary patterns have been beneficial in treating diabetes and the concomitant increase in AGEs. Glycaemic control is most frequently measured *via* fasting blood glucose or HbA1C concentrations.[131,132] However, it would be expected that these same glycaemic lowering effects from dietary changes alone would lead to reduced levels of AGE precursors in the circulation and in tissues as has been demonstrated by pharmaceutical treatments of diabetes.[133]

The substantial array of natural plant-derived compounds under investigation and development, offer unique opportunities for food formulation scientists to develop a range of functional food products that incorporate natural complexes which may contribute to the amelioration of AGE-related damage to organ systems.[134] The following section gives a brief overview of some newly identified plant-based compounds that have shown promise in the alleviation of AGEs in the body.

Aydın *et al.*[135] have shown that the antioxidative and antiglycating properties of carnosine can significantly reduce OS due to a galactose-induced ageing mechanism caused by AGEs. Carnosine is a dipeptide molecule formed from the amino acids beta-alanine and histidine and is naturally found at high levels in muscle and brain tissue. This molecule has been shown to have strong reactive oxygen species (ROS) scavenging capability as well as the ability to block unsaturated aldehydes formed *via* peroxidation of cell membrane fatty acids.

Guzmán and colleagues, have shown that the marine extract scalarin, inhibits RAGE in pancreatic cancer cell lines.[136] RAGE has recently emerged as a chemotherapeutic target for some cancers, as it is an important regulator of inflammation, stress and survival pathways. Dysregulation of these pathways may lead to carcinogenesis, resistance to chemotherapy and high metastatic potential, which are hallmarks of pancreatic cancer. The active compound derived from a marine sponge identified as *Euryspongia* cf. *rosea.*, scalarin is a sesterterpene natural product. Scalarin reduces the levels of RAGE and inhibits autophagy in pancreatic cancer cell lines. The researchers found that the IC_{50} for cytotoxicity ranged between 20 and 30 μM. The IC_{50}, is half maximal inhibitory concentration and is a measure of the potency of a compound in inhibiting a specific biological or biochemical function. Inhibition of autophagy limits tumour growth and tumorigenesis in pancreatic cancer, making scalarin a potential natural compound for treatment regimes.

An interesting study by Yu and co-workers,[137] examined pterostilbene which is a natural compound derived from blueberries. Their study looked at the anticancer, anti-inflammatory and antioxidative effects in a cell-based study. The pterostilbene compound significantly suppressed ROS production and reduced carbonyl levels of proteins. This compound also downregulated gene expression of inflammatory mediators including tumour necrosis factor alpha (TNF-α), IL-1β and IL-6. Further analysis indicated that pterostilbene inhibited AGE-induced RAGE up-regulation. Furthermore, it reduced nuclear factor kappa-light-chain-enhancer activated B cell (NF-κB) phosphorylation and MAPK activation, including p38 and ERK1/2 pathways induced by AGEs. These results indicate that pterostilbene might act as a promising agent to alleviate AGEs-induced OS and inflammation *via* the RAGE/MAPK/NF-κB pathway.

Using newer technologies, Jha and Ali, looked at methods for early detection and effective prevention of AGE-derived compounds.[138] Recently, nanotechnology has emerged as a reliable and sensitive technique for the detection and use of plant-mediated nanoparticles as sensors of glucose, glycated proteins, protein oxidation and AGEs. The rapid detection of AGE-related compounds could significantly aid in the implementation of diagnostic strategies for increased AGE-related stress in the body.

A further study by Sowndhar *et al.* recently examined diet derived AGEs (dAGE) from mice on a high AGE-induced diet. This study showed a significant rise in systemic inflammation, pro-inflammatory cytokines and chemokines and weight gain in the animal study. The research used a natural antioxidant co-treatment of curcumin and gallic acid that normalised dAGE circulatory levels and neutralised dAGE-induced systemic inflammation.[139]

Interestingly, a recent study showed that salidroside (rhodioloside), derived from the plant *Rhodiola rosea*, has a putative antidepressant effect and has been in use as a dietary supplement for many years. A study by Zhang and co-workers, examined this compound for its effect on diabetes-induced vascular endothelial dysfunction. The study found that salidroside restored nitric oxide levels with a concomitant relaxation of arterial stiffness and indicated a protective effect against high levels of AGEs.[140]

The research field over recent years has been highly active in the search for natural plant-based compounds that can reduce postprandial release of glucose into the bloodstream. Two important enzymes are involved in the postprandial release of glucose, α-amylase and α-glucosidase, and a recent study by Agawane *et al.*,[141] was aimed at reducing the levels of these enzymes. *Mentha arvensis* L. (wild mint) is a traditional plant-based Ayurveda agent that has been in use for centuries across the globe for its high level of menthol-related chemicals. Use of the extract from this plant in the study exhibited high ROS scavenging activity as well as a reduction in α-amylase levels by 50% and α-glucosidase by 68%. The research study also showed a remarkable decrease in AGE formation of 90%.[141] *Momordica charantia* extracted from bitter melon has also been shown to significantly reduce AGE levels and also promotes wound healing in diabetic patients.[142] *Nelumbo nucifera* leaf extract was evaluated by Zhang and colleagues for its potential to reduce the formation of AGEs. This study found a significant reduction of ROS and an inhibitory effect on the formation of AGEs by attenuating the conformational change in proteins induced by glycation.[143]

Another study that investigated the restriction of AGE formation was undertaken by Rajpurohit *et al.*[144] using lutein derived from *Tagetes erecta* (Mexican marigold). A seven-day exposure to this extract indicated a notable

AGE inhibition of approximately 89%. The non-enzymatic adduct formation between the keto group of sugar and amino group of proteins is one of the molecular-based causes of diabetic complications in the hyperglycaemic state. Inhibition of this process will be useful in the management of diabetic complications. Lutein has shown a positive dose-dependent inhibitory effect on protein glycation.[144]

Secondary metabolites from various species of lichen are known to be potent inhibitors of AGE formation and have additional positive vasodilative effects.[145] Glycation and production of AGEs are also known to damage skin, especially facial skin as this is continuously exposed to light. Gromwell root (*Lithospermum erythrorhizon*) (GR), has an extensive traditional medicinal history that includes providing multiple skin benefits. Glynn *et al.*[146] (2018), examined the effect of the plant extract shikonin and found it suppressed the inflammatory response by enhancing UV protection. The research also indicated that the extract dosage independently inhibited glycation and enhanced oxidative defences through nuclear factor erythroid 2-related factor 2 (Nrf2)/antioxidant response element activation. Inflammatory targets, NFκB and TNFα, were suppressed by GR extract and shikonin. Glyoxalase 1 (GLO1) and glutathione synthesis genes were additionally upregulated by GR extract and shikonin. GR extract boosted higher wavelength UV absorption in selected cosmetic matrices. The rationale for the use of GR extract and shikonin was supported by this research. By inhibiting glycation, modulating OS, suppressing inflammation and enhanced UV-absorptive properties, GR extract and shikonin potentially offer multiple skin benefits, either as a dietary supplement or for inclusion in a range of cosmetic skin products.[146]

In an older study by Uribarri *et al.*,[147] a cohort of 172 young (<45 years old) and older (>60 years old), healthy individuals were examined to determine whether the concentrations of specific serum AGE markers (CML or MG derivatives) were higher in older compared with younger persons and whether, independent of age, they correlated with the intake of dietary AGEs, as well as with circulating markers of OS and inflammation.[147] The results of this study found that circulating indicators of AGEs (CML and MG derivatives), although elevated in older participants, correlated with indicators of inflammation and OS across all ages. Indicators of both AGEs and OS are directly influenced by the intake of dietary AGEs, independent of age or energy intake.

A range of herbs and spices are a good source of complex polyphenols and have over recent years shown they could be strong inhibitors of AGE formation.

Antioxidant herbs and spices have been shown to inhibit tissue OS across a wide range of animal models and in human studies.[61,148–150] Many herbs and spices have been used for centuries in traditional medicine across the world. Prolonged oxidative stress can increase protein glycation in a range of tissue systems in the body and some herbs and spices contain potent antiglycation compounds. AGEs not only cause overproduction of free radical molecules but can also change intracellular signalling and alter gene expression, resulting in a release of pro-inflammatory cytokines leading to damage to organ systems. There is a growing list of beneficial herbs and spices that may reduce the effects of AGEs, as they have anti-glycation properties which are correlated to the total content of polyphenols. These antiglycation herbs and spices include:[151] *Allium sativum; Zingiber officinale; Thymus vulgaris; Petroselinum crissum; Murraya koenigli Spreng; Mentha piperita* L.; *Curcuma longa* L.; *Allium cepa* L.; *Allium fistulosum* and *Coriandrum sativum* L.

Starowicz and Zieliński (2019), examined 14 herbs and spices for their ability to reduce the formation of AGE-related toxicants.[61] The most potent glycation inhibitors, according to the BSA-MGO assay, were star anise (88%), cinnamon (85%), allspice (81%), and cloves (79%), whereas in the BSA–glucose measurement, oregano was noted to be a very effective inhibitor of the glycation process. Their research showed a high antiglycation ability of cinnamon, cloves and allspice. This range of herbs and spices are suggested to be potent inhibitors of AGE formation and ideal candidates for inclusion in functional food products.[134]

As a final example of novel compounds that can ameliorate AGE formation and reduce disease burden, Li *et al.* examined two novel flavoalkaloids, (−)-6-(5‴-*S*)-*N*-ethly-2-pyrrolidnone-epigallocatechin-*O*-gallate and (−)-6-(5‴-*R*)-*N*-ethyl-2-pyrrolidinone-epigallocatechin-*O*-gallate, both derived from white tea (*Camellia sinensis*). Results indicated a significant decrease in AGE formation and a reduction in inflammatory responses.[152]

Studies on the antidiabetic mechanisms of medicinal plants have shown that a high proportion of them induce hypoglycaemic activity by stimulating insulin secretion, augment peroxisome proliferator-activated receptors (PPARs) and inhibit α-amylase or α-glucosidase, glucagon-like peptide-1 (GLP-1) secretion and AGE formation.[153–162] Furthermore, plant extracts have the ability to increase free radical scavenging activity, plus antioxidant activity against reactive oxygen or nitrogen species (ROS/RNS) and contribute to preventing the development of a range of inflammation-induced disease conditions. Strategies to reduce AGE products in general are aimed at inhibiting glycation by several mechanisms. Plant extracts have been shown to inhibit the formation of the initial Schiff base and Amadori products. Other natural compounds inhibit AGE-formation products in the advanced phase. Some natural plant compounds blocked the aggregation of AGEs and have exhibited antioxidant activity and reduced AGE formation by preventing oxidation of Amadori products and metal-catalysed glucoxidation.

8.7 Conclusions and Recommendations

In recent years, the effects of AGEs on health and disease progression have received increasing attention. High AGE levels in the body correlate with the progression of many diseases, such as diabetes, CVD, kidney disease, neurodegeneration and potentially cancer.[32]

AGEs are a complex and heterogeneous group of compounds that have been implicated in age-related diseases and related complications. Circulating indicators of AGEs (CML and MG derivatives), although elevated in older population groups, correlate with indicators of inflammation and oxidative stress across all ages. Indicators of both AGEs and oxidative stress are directly influenced by the intake of dietary AGEs, independent of age or energy intake. Therefore, reduced consumption of these oxidants *via* dietary means may prove a safe strategy to prevent the exacerbation of age-related diseases, given that the ageing population is rising.

Many disease states and progression are linked to oxidative stress which is known to be part of the normal ageing process.[32,163–165] Additionally, this increased level of oxidative stress could also contribute to the glycation of lipids and proteins from heat-treated foods in the diet, forming high levels of AGEs. The Western diet is high in AGEs that have been found to be pro-inflammatory. Industrial scale processed food in Western culture is not only high in fat, proteins and sugars but often requires heating at high temperatures. This combination has been found to significantly contribute to the production and accumulation of toxic AGEs in the body.

Since AGEs have been shown to generate reactive oxygen intermediates, linkage of AGEs to the cell surface by their receptors concentrates oxidant stress on cellular targets, resulting in changes in gene expression and the cellular phenotype. The European Food Safety Authority (EFSA), in 2018 identified research priorities under the theme: Protecting consumers' health with independent scientific advice on the food chain. The full EFSA Report can be found at this link: (EFSA Strategy 2020 – Trusted science for safe food): https://www.efsa.europa.eu/sites/default/files/corporate_publications/files/strategy2020.pdf.

AGEs and a reduction of ageing-related diseases have been highlighted as a prominent area for possible future legislative measures, as currently there is no AGE-related legislation, although EFSA do have AGE levels in food on their watchlist. Despite their complexity and widespread pathological distribution, the formation of AGEs (from all sources) promote the same chemical and biological outcomes. These are, the formation of covalent cross-links between proteins and an increase in the inflammatory response. These mechanisms are proposed as the central underlying process that leads to organ damage. The substantial body of research has sparked an intense debate on the risk of dietary AGEs to human health. In some studies analysis has provided evidence that dietary AGEs contribute to clinical outcomes and disease. Furthermore, human studies have failed to observe any validated measurable biomarker association. Whether AGEs are a cause of these debilitating diseases through environmental and food sources or represent associated symptoms of these diseases still needs to be determined by additional comprehensive research. Recently, many researchers have begun to investigate the effects of AGE intake-induced variations of gut microbiota on disease progression. These microbiome studies may eventually explain the role of the gut microbiota and the impact of AGEs on health.[166–171]

Gaps in analytical measurement techniques remain, especially the availability of rapid, affordable and reliable alternative methods with readily available standards and validated methods.[79] It is clear that difficulties relating to current methods for the detection and measurement of AGEs are of particular concern, including the lack of a universally established method of detection or a defined unit of measurement.[79] Therapeutic agents used for the treatment of AGE accumulation are an active area of research and newly discovered compounds are robustly being investigated. However, it remains controversial as to whether AGEs are the cause, or the consequence of the complications observed in a range of disease conditions. Understanding the pathogenic mechanisms of AGEs is paramount to the development of strategies against oxidative stress, diabetic and cardiovascular complications in particular.

Treatment strategies show medicinal plants are effective in diabetic complications by reducing the formation of AGEs, and many of them also possess a variety of beneficial effects on other pathological conditions, including the complications of inflammatory responses, which are fundamentally responsible for a wide range of disease states. Such plant derived bioactive compounds may be appropriate alternatives or adjuncts to available antidiabetic and other pharmacological intervention.

Inflammation is a critical step for the development of many complex diseases and disorders including autoimmune diseases, metabolic syndrome, neurodegenerative diseases, cancers and cardiovascular diseases.[172] Uncontrolled chronic inflammation may well be the aetiological cause of many of the disease states of ageing. An alternative strategy to those described above is to enhance the naturally occurring detoxification pathways for AGEs and AGE precursors. As previously discussed, the glyoxalase system is the primary mechanism for detoxifying reactive dicarbonyls that are relevant in a multitude of disease states.[45,173] Aging and degenerative diseases are associated with reduced levels of GLO1 and reduced levels of the critical cofactor glutathione.[174,175] Therefore, compounds that increase GLO1 activity, either through increased transcription or post-translationally, or that increase levels of the reduced form of glutathione have the potential to reduce *de novo* formation of AGEs.[176,177]

In addition to other recommendations, it is clear that cooking methods play a large role in the formation of AGEs and related compounds. Currently, it is recommended by many health organisations (American Heart Association; Diabetes UK; Cancer Research UK), that direct-heat methods of food preparation both on an industrial scale and in the home, should be limited.[118] Alternative cooking methods such as stewing, boiling, steaming and poaching or other slow cooking methods are therefore recommended. These methods of food preparation do not generate high levels of these dangerous and pathological AGEs and related compounds.

Understanding the mechanism of formation of AGEs and the effects of digestion and metabolism can progress approaches to limit the risk and further develop mitigation stratagems of these process or diet-induced contaminants. It is clear that strategies should be developed that restrict the initial formation of AGEs, although some naturally derived plant extracts show that they can also reduce levels of AGEs once formed in the body. In considering a treatment regime for complex multifactorial diseases associated with AGE accumulation, it is likely that a multifaceted approach will be most effective. Diet, nutrition and environmental factors and modifications of these should be considered and implemented.

8.8 Glossary of Terms

AGE	Advanced glycation endproducts
ALE	Advanced lipoxidation endproducts
AP	Amadori product
ALT-711	Alagebrium
BSA	Bovine serum albumin
BSA-MGO	Bovine serum albumin-methylglyoxal
CEL	N^{ε}-1-Carboxyethyllysine
CML	N^{ε}-Carboxymethyllysine
CVD	Cardiovascular disease
dAGE	Dietary advanced glycation endproducts
DC	1,2-Dicarbonyl compounds
DPTC	4,5-Dimethyl-3-phenacylthiazolium chloride
EC	Endothelial cells
ED	Endocrine disruptors
ELISA	Enzyme-linked immunosorbent assay
ERK1/2	Extracellular signal-regulated kinases
fM	Femtomole concentration
GLO-1	Glyoxalase 1
GLP-1	Glucagon-like peptide-1
GR	Gromwell root
HAA	Heterocyclic aromatic amines
HbA1C	A measure of the beta-*N*-1-deoxy fructosyl component of haemoglobin
HMF	Hydroxymethylfurfural
HP	Heyns product
HPLC	High-performance liquid chromatography

HPLC-ITMS/MS	High-performance-liquid chromatography-ion trap mass spectrometry/mass spectrometry
IC_{50}	Half maximal inhibitory concentration
IL-1β	Interleukin 1 beta
IL-6	Interleukin 6
K_d	Quantitative measurement of antibody affinity
MAPK	Mitogen-activated protein kinase
MG	Methylglyoxal
MG-H1	N^{δ}-(5-Hydro-5-methyl-4-imidazolon-2-yl)-ornithine
MRP	Maillard reaction products
NF-κB	Nuclear factor kappa-light-chain-enhancer of activated B cells
TNF-α	Tumour necrosis factor alpha
nM	Nanomole concentration
Nrf2	Nuclear factor erythroid 2-related factor 2
OS	Oxidative stress
PAH	Polycyclic aromatic hydrocarbons
PPAR	Peroxisome proliferator-activated receptors
ppb	Parts per billion
p38	Mitogen-activated protein kinase
RAGE	Receptor for advanced glycation endproducts
RNS	Reactive nitrogen species
ROS	Reactive oxygen species
ss DNA	Single-stranded DNA
ss RNA	Single-stranded RNA
SELEX	Systematic evolution of ligands by exponential enrichment
TLC	Thin-layer chromatography
UHPLC-MS/MS	Ultra-high-performance liquid chromatography-mass spectrometry/mass spectrometry
Vitamin B_6	Pyridoxal 5′-phosphate

References

1. J. Uribarri, S. Woodruff, S. Goodman, *et al.*, Advanced Glycation End Products in Foods and a Practical Guide to Their Reduction in the Diet, *J. Am. Diet. Assoc.*, 2010, **110**(6), 911–916.e12.
2. M. Gehringer and S. A. Laufer, Emerging and Re-Emerging Warheads for Targeted Covalent Inhibitors: Applications in Medicinal Chemistry and Chemical Biology, *J. Med. Chem.*, 2019, **62**, 5673–5724.
3. N. Aljahdali and F. Carbonero, Impact of Maillard reaction products on nutrition and health: Current knowledge and need to understand their fate in the human digestive system, *Crit. Rev. Food Sci. Nutr.*, 2019, **59**(3), 474–487.
4. J. K. Román and J. J. Wilker, Cooking Chemistry Transforms Proteins into High-Strength Adhesives, *J. Am. Chem. Soc.*, 2019, **141**(3), 1359–1365.

5. O. G. Andriotis, K. Elsayad, D. E. Smart, M. Nalbach, D. E. Davies and P. J. Thurner, Hydration and nanomechanical changes in collagen fibrils bearing advanced glycation end-products, *Biomed. Opt. Express*, 2019, **10**(4), 1841–1855.
6. S. Rowan, E. Bejarano and A. Taylor, Mechanistic targeting of advanced glycation end-products in age-related diseases, *Biochim. Biophys. Acta, Mol. Basis Dis.*, 2018, **1864**(12), 3631–3643.
7. E. Bejarano and A. Taylor, Too sweet: Problems of protein glycation in the eye, *Exp. Eye Res.*, 2019, **178**, 255–262.
8. C.-T. Li, S.-C. How, M.-E. Chen, *et al.*, Effects of glycation on human γd-crystallin proteins by different glycation-inducing agents, *Int. J. Biol. Macromol.*, 2018, **118**, 442–451.
9. M. Kosmopoulos, D. Drekolias, P. D. Zavras, C. Piperi and A. G. Papavassiliou, Impact of advanced glycation end products (AGEs) signaling in coronary artery disease, *Biochim. Biophys. Acta, Mol. Basis Dis.*, 2019, **1865**(3), 611–619.
10. D. Ngai, M. Lino and M. P. Bendeck, Cell-Matrix Interactions and Matricrine Signaling in the Pathogenesis of Vascular Calcification, *Front. Cardiovasc. Med.*, 2018, **5**, 174.
11. M. Meurillon and E. Engel, Mitigation strategies to reduce the impact of heterocyclic aromatic amines in proteinaceous foods, *Trends Food Sci. Technol.*, 2016, **50**, 70–84.
12. B. A. Mogol and V. Gökmen, Thermal process contaminants: acrylamide, chloropropanols and furan, *Curr. Opin. Food Sci.*, 2016, **7**, 86–92.
13. C. Nerín, M. Aznar and D. Carrizo, Food contamination during food process, *Trends Food Sci. Technol.*, 2016, **48**, 63–68.
14. M. Nooshkam, M. Varidi and M. Bashash, The Maillard reaction products as food-born antioxidant and antibrowning agents in model and real food systems, *Food Chem.*, 2019, **275**, 644–660.
15. M. Karbasi and A. Madadlou, Interface-related attributes of the Maillard reaction-born glycoproteins, *Crit. Rev. Food Sci. Nutr.*, 2018, **58**(10), 1595–1603.
16. A. M. Băbţan, A. Ilea, B. A. Boşca, *et al.*, Advanced glycation end products as biomarkers in systemic diseases: premises and perspectives of salivary advanced glycation end products, *Biomarkers Med.*, 2019, **13**(6), 479–495.
17. K. Nowotny, D. Schröter, M. Schreiner and T. Grune, Dietary advanced glycation end products and their relevance for human health, *Ageing Res. Rev.*, 2018, **47**, 55–66.
18. J. Koska, A. Saremi, S. Howell, *et al.*, Advanced Glycation End Products, Oxidation Products, and Incident Cardiovascular Events in Patients With Type 2 Diabetes, *Diabetes Care*, 2018, **41**(3), 570.
19. S.-i. Yamagishi, Diabetes and Advanced Glycation End Products, in *Diabetes and Aging-related Complications*, ed. S.-i. Yamagishi, Springer, Singapore, 2018, pp. 201–212.

20. N. Rabbani and P. J. Thornalley, Advanced glycation end products in the pathogenesis of chronic kidney disease, *Kidney Int.*, 2018, **93**(4), 803–813.
21. S. Y. Rhee and Y. S. Kim, The Role of Advanced Glycation End Products in Diabetic Vascular Complications, *Diabetes Metab. J.*, 2018, **42**(3), 188–195.
22. C. Delgado-Andrade and V. Fogliano, Dietary Advanced Glycosylation End-Products (dAGEs) and Melanoidins Formed through the Maillard Reaction: Physiological Consequences of their Intake, *Annu. Rev. Food Sci. Technol.*, 2018, **9**(1), 271–291.
23. A. Bellahcène, M.-J. Nokin, V. Castronovo and C. Schalkwijk, Methylglyoxal-derived stress: An emerging biological factor involved in the onset and progression of cancer, *Semin. Cancer Biol.*, 2018, **49**, 64–74.
24. J. Bellier, M.-J. Nokin, E. Lardé, *et al.*, Methylglyoxal, a potent inducer of AGEs, connects between diabetes and cancer, *Diabetes Res. Clin. Pract.*, 2019, **148**, 200–211.
25. C. Miglio, E. Chiavaro, A. Visconti, V. Fogliano and N. Pellegrini, Effects of Different Cooking Methods on Nutritional and Physicochemical Characteristics of Selected Vegetables, *J. Agric. Food Chem.*, 2008, **56**(1), 139–147.
26. J. L. J. M. Scheijen, E. Clevers, L. Engelen, *et al.*, Analysis of advanced glycation endproducts in selected food items by ultra-performance liquid chromatography tandem mass spectrometry: Presentation of a dietary AGE database, *Food Chem.*, 2016, **190**, 1145–1150.
27. M. Shaikh, S. Siddiqui, H. Zafar, *et al.*, Antiglycation activity of triazole Schiff's bases against fructose-mediated glycation: *In vitro* and *in silico* study, *Med. Chem.*, 2019, DOI: 10.2174/1573406415666190212105718.
28. N. M. Sjoblom, M. M. G. Kelsey and R. A. Scheck, A Systematic Study of Selective Protein Glycation, *Angew. Chem., Int. Ed.*, 2018, **57**(49), 16077–16082.
29. A. D. Troise, M. Wiltafsky, V. Fogliano and P. Vitaglione, The quantification of free Amadori compounds and amino acids allows to model the bound Maillard reaction products formation in soybean products, *Food Chem.*, 2018, **247**, 29–38.
30. H. Prasch, C. Hojnik, T. K. Lindhorst, B. Didak, L. Landemarre and T. M. Wrodnigg, New lectin ligands: Testing of Amadori rearrangement products with a series of mannoside-specific lectins, *Carbohydr. Res.*, 2019, **475**, 65–68.
31. S. D. Sharma and M. Barone, Deleterious Consequences of Dietary Advanced Glycation End Products on Human Health Due to Oxidative Stress and Inflammation, *Dietary Patterns, Food Chemistry and Human Health*, Springer International Publishing, Cham, 2019, pp. 1–13.
32. J.-A. Lin, C.-H. Wu and G.-C. Yen, Perspective of Advanced Glycation End Products on Human Health, *J. Agric. Food Chem.*, 2018, **66**(9), 2065–2070.

33. A. Stirban, T. Gawlowski and M. Roden, Vascular effects of advanced glycation endproducts: Clinical effects and molecular mechanisms, *Mol. Metab.*, 2014, **3**(2), 94–108.
34. M. Mol, G. Degani, C. Coppa, *et al.*, Advanced lipoxidation end products (ALEs) as RAGE binders: Mass spectrometric and computational studies to explain the reasons why, *Redox Biol.*, 2018, **2018**, 101083.
35. J. d. D. Alché, A concise appraisal of lipid oxidation and lipoxidation in higher plants, *Redox Biol.*, 2019, **2019**, 101136.
36. C. B. Afonso and C. M. Spickett, Lipoproteins as targets and markers of lipoxidation, *Redox Biol.*, 2018, **2018**, 101066.
37. E. Gianazza, M. Brioschi, A. M. Fernandez and C. Banfi, Lipoxidation in cardiovascular diseases, *Redox Biol.*, 2019, **2019**, 101119.
38. T. Baldensperger, T. Jost, A. Zipprich and M. A. Glomb, Novel α-Oxoamide Advanced-Glycation Endproducts within the N6-Carboxymethyl Lysine and N6-Carboxyethyl Lysine Reaction Cascades, *J. Agric. Food Chem.*, 2018, **66**(8), 1898–1906.
39. Y. Wang, H. Hu, D. J. McClements, *et al.*, Effect of fatty acids and triglycerides on the formation of lysine-derived advanced glycation endproducts in model systems exposed to frying temperature, *RSC Adv.*, 2019, **9**(27), 15162–15170.
40. S. Parisi and W. Luo, Maillard Reaction and Processed Foods—Main Chemical Products, in *Chemistry of Maillard Reactions in Processed Foods*, ed. S. Parisi and W. Luo, Springer International Publishing, Cham, 2018, pp. 53–59.
41. R. K. Singla, A. K. Dubey, S. M. Ameen, S. Montalto and S. Parisi, Analytical Methods for the Determination of Maillard Reaction Products in Foods. An Introduction, in *Analytical Methods for the Assessment of Maillard Reactions in Foods*, ed. R. K. Singla, A. K. Dubey, S. M. Ameen, S. Montalto and S. Parisi, Springer International Publishing, Cham, 2018, pp. 1–14.
42. J. Degen, M. Hellwig and T. Henle, 1,2-Dicarbonyl Compounds in Commonly Consumed Foods, *J. Agric. Food Chem.*, 2012, **60**(28), 7071–7079.
43. L. F. Bjeldanes and H. Chew, Mutagenicity of 1,2-dicarbonyl compounds: Maltol, kojic acid, diacetyl and related substances, *Mutat. Res., Genet. Toxicol.*, 1979, **67**(4), 367–371.
44. T. Hirayama, N. Yamada, M. Nohara and S. Fukui, The existence of the 1,2-dicarbonyl compounds glyoxal, methyl glyoxal and diacetyl in autoxidised edible oils, *J. Sci. Food Agric.*, 1984, **35**(12), 1357–1362.
45. M. Hellwig, S. Gensberger-Reigl, T. Henle and M. Pischetsrieder, Food-derived 1,2-dicarbonyl compounds and their role in diseases, *Semin. Cancer Biol.*, 2018, **49**, 1–8.
46. E. D. Cömert and V. Gökmen, Kinetic evaluation of the reaction between methylglyoxal and certain scavenging compounds and determination of their *in vitro* dicarbonyl scavenging activity, *Food Res. Int.*, 2019, **121**, 257–268.

47. S. Treibmann, F. Spengler, J. Degen, J. Löbner and T. Henle, Studies on the Formation of 3-Deoxyglucosone- and Methylglyoxal-Derived Hydroimidazolones of Creatine during Heat Treatment of Meat, *J. Agric. Food Chem.*, 2019, **67**(20), 5874–5881.
48. M. Aragno and R. Mastrocola, Dietary Sugars and Endogenous Formation of Advanced Glycation Endproducts: Emerging Mechanisms of Disease, *Nutrients*, 2017, **9**(4), 385.
49. L. J. Sparvero, D. Asafu-Adjei, R. Kang, *et al.*, RAGE (Receptor for Advanced Glycation Endproducts), RAGE Ligands, and their role in Cancer and Inflammation, *J. Transl. Med.*, 2009, **7**(1), 17.
50. T. Chavakis, A. Bierhaus and P. P. Nawroth, RAGE (receptor for advanced glycation end products): a central player in the inflammatory response, *Microbes Infect.*, 2004, **6**(13), 1219–1225.
51. C. A. Downs, N. M. Johnson, G. Tsaprailis and M. N. Helms, RAGE-induced changes in the proteome of alveolar epithelial cells, *J. Proteomics*, 2018, **177**, 11–20.
52. J. Derk, M. MacLean, J. Juranek and A. M. Schmidt, The Receptor for Advanced Glycation Endproducts (RAGE) and Mediation of Inflammatory Neurodegeneration, *J. Alzheimer's Dis. Parkinsonism*, 2018, **8**(1), 421.
53. K. Batkulwar, R. Godbole, R. Banarjee, O. Kassaar, R. J. Williams and M. J. Kulkarni, Advanced Glycation End Products Modulate Amyloidogenic APP Processing and Tau Phosphorylation: A Mechanistic Link between Glycation and the Development of Alzheimer's Disease, *ACS Chem. Neurosci.*, 2018, **9**(5), 988–1000.
54. Z. S. Ataç, M. Alaylıoğlu, E. Dursun, D. Gezen-Ak, S. Yılmazer and H. Gürvit, G82S polymorphism of receptor for advanced glycation end products gene and serum soluble RAGE levels in mild cognitive impairment and dementia of Alzheimer's type patients in Turkish population, *J. Clin. Neurosci.*, 2019, **59**, 197–201.
55. M. MacLean, J. Derk, H. H. Ruiz, J. K. Juranek, R. Ramasamy and A. M. Schmidt, The Receptor for Advanced Glycation End Products (RAGE) and DIAPH1: Implications for vascular and neuroinflammatory dysfunction in disorders of the central nervous system, *Neurochem. Int.*, 2019, **126**, 154–164.
56. Q. Zhu and E. A. Smith, Diaphanous-1 affects the nanoscale clustering and lateral diffusion of receptor for advanced glycation endproducts (RAGE), *Biochim. Biophys. Acta, Mol. Basis Dis.*, 2019, **1861**(1), 43–49.
57. M. A. Incalza, R. D'Oria, A. Natalicchio, S. Perrini, L. Laviola and F. Giorgino, Oxidative stress and reactive oxygen species in endothelial dysfunction associated with cardiovascular and metabolic diseases, *Vasc. Pharmacol.*, 2018, **100**, 1–19.
58. B. I. Hudson and M. E. Lippman, Targeting RAGE Signaling in Inflammatory Disease, *Annu. Rev. Med.*, 2018, **69**(1), 349–364.
59. A. Rojas, M. Morales, I. Gonzalez and P. Araya, Inhibition of RAGE Axis Signaling: A Pharmacological Challenge, *Curr. Drug Targets*, 2019, **20**(3), 340–346.

60. Y. Zhang, X. Yang, C. Qiu, F. Liu, P. Liu and Z. Liu, Matrine suppresses AGE-induced HAEC injury by inhibiting ROS-mediated NRLP3 inflammasome activation, *Eur. J. Pharmacol.*, 2018, **822**, 207–211.
61. M. Starowicz and H. Zieliński, Inhibition of Advanced Glycation End-Product Formation by High Antioxidant-Leveled Spices Commonly Used in European Cuisine, *Antioxidants*, 2019, **8**(4), 100.
62. A. D. Troise, Analytical strategies to depict the fate of the Maillard reaction in foods, *Curr. Opin. Food Sci.*, 2018, **19**, 15–22.
63. A. D. Troise, C. C. Berton-Carabin and V. Fogliano, Amadori products formation in emulsified systems, *Food Chem.*, 2016, **199**, 51–58.
64. T. Goldberg, W. Cai, M. Peppa, *et al.*, Advanced glycoxidation end products in commonly consumed foods, *J. Am. Diet. Assoc.*, 2004, **104**(8), 1287–1291.
65. G. L. J. Hull, J. V. Woodside, J. M. Ames and G. J. Cuskelly, N^{ε}-(carboxymethyl)lysine content of foods commonly consumed in a Western style diet, *Food Chem.*, 2012, **131**(1), 170–174.
66. J. L. J. M. Scheijen, N. M. J. Hanssen, M. M. van Greevenbroek, *et al.*, Dietary intake of advanced glycation endproducts is associated with higher levels of advanced glycation endproducts in plasma and urine: The CODAM study, *Clin. Nutr.*, 2018, **37**(3), 919–925.
67. C. Niquet-Léridon, P. Jacolot, C.-N. Niamba, N. Grossin, E. Boulanger and F. J. Tessier, The rehabilitation of raw and brown butters by the measurement of two of the major Maillard products, N^{ε}-carboxymethyllysine and 5-hydroxymethylfurfural, with validated chromatographic methods, *Food Chem.*, 2015, **177**, 361–368.
68. A. Gómez-Ojeda, S. Jaramillo-Ortíz, K. Wrobel, *et al.*, Comparative evaluation of three different ELISA assays and HPLC-ESI-ITMS/MS for the analysis of N^{ε}-carboxymethyl lysine in food samples, *Food Chem.*, 2018, **243**, 11–18.
69. Y. Higashimoto, T. Matsui, Y. Nishino, *et al.*, Blockade by phosphorothioate aptamers of advanced glycation end products-induced damage in cultured pericytes and endothelial cells, *Microvasc. Res.*, 2013, **90**, 64–70.
70. S. Maeda, T. Matsui, A. Ojima, *et al.*, DNA Aptamer Raised against Advanced Glycation End Products Prevents Abnormalities in Electroretinograms of Experimental Diabetic Retinopathy, *Ophthalmic Res.*, 2015, **54**(4), 175–180.
71. A. Ojima, T. Matsui, S. Maeda, *et al.*, DNA aptamer raised against advanced glycation end products inhibits melanoma growth in nude mice, *Lab. Invest.*, 2014, **94**, 422.
72. S.-i. Yamagishi, K. Taguchi and K. Fukami, DNA-aptamers raised against AGEs as a blocker of various aging-related disorders, *Glycoconjugate J.*, 2016, **33**(4), 683–690.
73. R. M. Kärkkäinen, M. R. Drasbek, I. McDowall, C. J. Smith, N. W. G. Young and G. A. Bonwick, Aptamers for safety and quality assurance in the food industry: detection of pathogens, *Int. J. Food Sci. Technol.*, 2011, **46**(3), 445–454.

74. R. Rapini and G. Marrazza, Electrochemical aptasensors for contaminants detection in food and environment: Recent advances, *Bioelectrochemistry*, 2017, **118**, 47–61.
75. F. Arduini, L. Micheli, D. Moscone, *et al.*, Electrochemical biosensors based on nanomodified screen-printed electrodes: Recent applications in clinical analysis, *TrAC, Trends Anal. Chem.*, 2016, **79**, 114–126.
76. J.-M. Liu, Y. Hu, Y.-K. Yang, *et al.*, Emerging functional nanomaterials for the detection of food contaminants, *Trends Food Sci. Technol.*, 2018, **71**, 94–106.
77. T. Matsui, Y. Higashimoto, Y. Nishino, N. Nakamura, K. Fukami and S.-i. Yamagishi, RAGE-Aptamer Blocks the Development and Progression of Experimental Diabetic Nephropathy, *Diabetes*, 2017, **66**(6), 1683.
78. T. R. Olsen, X. Meng, M. Stojanovic and Q. Lin, Towards Precision Medicine: Rapid Microfluidic Isolation of Aptamers for Glycan Targets, *ECS Meeting Abstracts*, 2019, MA2019-01(44), p. 2121.
79. R. H. Stadler, Foreword for Food Processing-Derived Contaminants in Food Analysis, *Reference Module in Food Science*, Elsevier, 2016.
80. M. Dominguez-Villar and D. A. Hafler, Regulatory T cells in autoimmune disease, *Nat. Immunol.*, 2018, **19**(7), 665–673.
81. P. Reddy, J. Leong and I. Jialal, Amino acid levels in nascent metabolic syndrome: A contributor to the pro-inflammatory burden, *J. Diabetes Complicat.*, 2018, **32**(5), 465–469.
82. J. Stephenson, E. Nutma, P. van der Valk and S. Amor, Inflammation in CNS neurodegenerative diseases, *Immunology*, 2018, **154**(2), 204–219.
83. C. Spagnuolo, S. Moccia and G. L. Russo, Anti-inflammatory effects of flavonoids in neurodegenerative disorders, *Eur. J. Med. Chem.*, 2018, **153**, 105–115.
84. J. Todoric and M. Karin, The Fire within: Cell-Autonomous Mechanisms in Inflammation-Driven Cancer, *Cancer Cell*, 2019, **35**(5), 714–720.
85. A. Annibaldi and P. Meier, Checkpoints in TNF-Induced Cell Death: Implications in Inflammation and Cancer, *Trends Mol. Med.*, 2018, **24**(1), 49–65.
86. B. Ritter and F. R. Greten, Modulating inflammation for cancer therapy, *J. Exp. Med.*, 2019, **16**(6), 1234–1243.
87. K. Taniguchi and M. Karin, NF-κB, inflammation, immunity and cancer: coming of age, *Nat. Rev. Immunol.*, 2018, **18**, 309.
88. N. Pechlivani and R. A. Ajjan, Thrombosis and Vascular Inflammation in Diabetes: Mechanisms and Potential Therapeutic Targets, *Front. Cardiovasc. Med.*, 2018, **5**(1), DOI: 10.3389/fcvm.2018.00001.
89. I. Ferreira, P. Hovind, C. G. Schalkwijk, H.-H. Parving, C. D. A. Stehouwer and P. Rossing, Biomarkers of inflammation and endothelial dysfunction as predictors of pulse pressure and incident hypertension in type 1 diabetes: a 20 year life-course study in an inception cohort, *Diabetologia*, 2018, **61**(1), 231–241.

90. T. Reinehr and C. L. Roth, Inflammation Markers in Type 2 Diabetes and the Metabolic Syndrome in the Pediatric Population, *Curr. Diabetes Rep.*, 2018, **18**(12), 131.
91. C. Herder and N. Hermanns, Subclinical inflammation and depressive symptoms in patients with type 1 and type 2 diabetes, *Semin. Immunopathol.*, 2019, **41**(4), 477–489.
92. M. I. Maiorino, G. Bellastella, D. Giugliano and K. Esposito, From inflammation to sexual dysfunctions: a journey through diabetes, obesity, and metabolic syndrome, *J. Endocrinol. Invest.*, 2018, **41**(11), 1249–1258.
93. L. Ferrucci and E. Fabbri, Inflammageing: chronic inflammation in ageing, cardiovascular disease, and frailty, *Nat. Rev. Cardiol.*, 2018, **15**(9), 505–522.
94. T. Ishida, M. Ishida, S. Tashiro and Y. Takeishi, DNA Damage and Senescence-Associated Inflammation in Cardiovascular Disease, *Biol. Pharm. Bull.*, 2019, **42**(4), 531–537.
95. S. Zhong, L. Li, X. Shen, *et al.*, An update on lipid oxidation and inflammation in cardiovascular diseases, *Free Radical Biol. Med.*, 2019, DOI: 10.1016/j.freeradbiomed.2019.03.036, in press.
96. Y. Deng, L. Song, X. Nie, W. Shou and X. Li, Prenatal inflammation exposure-programmed cardiovascular diseases and potential prevention, *Pharmacol. Ther.*, 2018, **190**, 159–172.
97. L. J. H. Rasmussen, T. E. Moffitt, J. Eugen-Olsen, *et al.*, Cumulative childhood risk is associated with a new measure of chronic inflammation in adulthood, *J. Child Psychol. Psychiatry*, 2019, **60**(2), 199–208.
98. C. Gabay and I. Kushner, Acute-Phase Proteins and Other Systemic Responses to Inflammation, *N. Engl. J. Med.*, 1999, **340**(6), 448–454.
99. A. Prigent, J. Gonzales, T. Durand, *et al.*, Acute inflammation downregulates alpha-synuclein expression in enteric neurons, *J. Neurochem.*, 2019, **148**(6), 746–760.
100. J. Kealy, C. Murray, E. W. Griffin, *et al.*, Acute inflammation alters energy metabolism in mice and humans: Role in sickness-induced hypoactivity, impaired cognition and delirium, *bioRxiv*, 2019, 642967.
101. E. C. Schroeder, A. D. Lane-Cordova, S. M. Ranadive, T. Baynard and B. Fernhall, Influence of fitness and age on the endothelial response to acute inflammation, *Exp. Physiol.*, 2018, **103**(6), 924–931.
102. T. Van der Lugt, A. R. Weseler, W. A. Gebbink, M. F. Vrolijk, A. Opperhuizen and A. Bast, Dietary Advanced Glycation Endproducts Induce an Inflammatory Response in Human Macrophages in Vitro, *Nutrients*, 2018, **10**(12), 1868.
103. J. Tekampe, H. van Middendorp, F. C. G. J. Sweep, S. H. P. P. Roerink, A. R. M. M. Hermus and A. W. M. Evers, Chapter Four – Human Pharmacological Conditioning of the Immune and Endocrine System: Challenges and Opportunities, in *International Review of Neurobiology*, ed. L. Colloca, Academic Press, 2018, vol. 138, pp. 61–80.

104. G. Ravichandran, D. K. Lakshmanan, K. Raju, *et al.*, Food advanced glycation end products as potential endocrine disruptors: An emerging threat to contemporary and future generation, *Environ. Int.*, 2019, **123**, 486–500.
105. T. Fiolet, B. Srour, L. Sellem, *et al.*, Consumption of ultra-processed foods and cancer risk: results from NutriNet-Santé prospective cohort, *BMJ*, 2018, **360**, k322.
106. D. Ruiz, M. Becerra, J. S. Jagai, K. Ard and R. M. Sargis, Disparities in Environmental Exposures to Endocrine-Disrupting Chemicals and Diabetes Risk in Vulnerable Populations, *Diabetes Care*, 2018, **41**(1), 193.
107. J. Yang, W. Song, X. Wang, *et al.*, Migration of phthalates from plastic packages to convenience foods and its cumulative health risk assessments, *Food Addit. Contam., Part B*, 2019, 1–8.
108. L. N. Vandenberg, Low dose effects challenge the evaluation of endocrine disrupting chemicals, *Trends Food Sci. Technol.*, 2019, **84**, 58–61.
109. A. Ernstoff, M. Niero, J. Muncke, *et al.*, Challenges of including human exposure to chemicals in food packaging as a new exposure pathway in life cycle impact assessment, *Int. J. Life Cycle Assess.*, 2019, **24**(3), 543–552.
110. A. M. Soto and C. Sonnenschein, Endocrine disruptors — putting the mechanistic cart before the phenomenological horse, *Nat. Rev. Endocrinol.*, 2018, **14**(6), 317–318.
111. L. Trasande and B. Blumberg, Endocrine Disruptors as Obesogens, in *Pediatric Obesity: Etiology, Pathogenesis and Treatment*, ed. M. S. Freemark, Springer International Publishing, Cham, 2018, pp. 243–253.
112. B. Werner, I. Norbert, B. Thomas, *et al.*, Occurrence of chemicals with known or suspected endocrine disrupting activity in drinking water, groundwater and surface water, Austria 2017/2018, *Die Bodenkultur J. Land Manage. Food and Environ.*, 2018, **69**(3), 155–173.
113. H. Wu, L.-H. Wu, F. Wang, C.-J. Gao, D. Chen and Y. Guo, Several environmental endocrine disruptors in beverages from South China: occurrence and human exposure, *Environ. Sci. Pollut. Res.*, 2019, **26**(6), 5873–5884.
114. K. Tamada, S. Nakajima, N. Ogawa, *et al.*, Papaverine identified as an inhibitor of high mobility group box 1/receptor for advanced glycation end-products interaction suppresses high mobility group box 1-mediated inflammatory responses, *Biochem. Biophys. Res. Commun.*, 2019, **511**(3), 665–670.
115. N. Adachi, I. Kanazawa, K.-i. Tanaka, *et al.*, Insulin-Like Growth Factor-I Protects Against the Detrimental Effects of Advanced Glycation End Products and High Glucose in Myoblastic C2C12 Cells, *Calcif. Tissue Int.*, 2019, **105**(1), 89–96.

116. A. K. Sharma, V. R. Sharma, G. K. Gupta, G. M. Ashraf and M. A. Kamal, Advanced Glycation End Products (AGEs), Glutathione and Breast Cancer: Factors, Mechanism and Therapeutic Interventions, *Curr. Drug Metab.*, 2019, **20**(1), 65–71.
117. R. Ghassem Zadeh and V. Yaylayan, Indole: A Promising Scavenging Agent for Methylglyoxal and Related Carbonyls in Tryptophan Containing Maillard Model Systems, *J. Agric. Food Chem.*, 2019, **67**(22), 6359–6365.
118. P. Jud and H. Sourij, Therapeutic options to reduce advanced glycation end products in patients with diabetes mellitus: A review, *Diabetes Res. Clin. Pract.*, 2019, **148**, 54–63.
119. F. A. Qais, M. S. Khan, A. S. Althubiani, S. B. Al-Ghamdi and I. Ahmad, Chapter 13-Understanding Biochemical and Molecular Mechanism of Complications of Glycation and Its Management by Herbal Medicine, in *New Look to Phytomedicine*, ed. M. S. Ahmad Khan, I. Ahmad and D. Chattopadhyay, Academic Press, 2019, pp. 331–366.
120. X. Wu, G. Zhang, X. Hu, J. Pan, Y. Liao and H. Ding, Inhibitory effect of epicatechin gallate on protein glycation, *Food Res. Int.*, 2019, **122**, 230–240.
121. S. Arfin, G. A. Siddiqui, A. Naeem and S. Moin, Inhibition of advanced glycation end products by isoferulic acid and its free radical scavenging capacity: An *in vitro* and molecular docking study, *Int. J. Biol. Macromol.*, 2018, **118**, 1479–1487.
122. H. Liu, C. Wang, X. Qi, J. Zou and Z. Sun, Antiglycation and antioxidant activities of mogroside extract from Siraitia grosvenorii (Swingle) fruits, *J. Food Sci. Technol.*, 2018, **55**(5), 1880–1888.
123. A. H. Burstein, M. Sabbagh, R. Andrews, C. Valcarce, I. Dunn and L. Altstiel, Development of Azeliragon, an Oral Small Molecule Antagonist of the Receptor for Advanced Glycation Endproducts, for the Potential Slowing of Loss of Cognition in Mild Alzheimer's Disease, *J. Prev. Alzheimer's Dis.*, 2018, **5**(2), 149–154.
124. C. Lei, S. Wu, C. Wen, *et al.*, Zafirlukast attenuates advanced glycation end-products (AGEs)-induced degradation of articular extracellular matrix (ECM), *Int. Immunopharmacol.*, 2019, **68**, 68–73.
125. S. V. Arnold, S. E. Inzucchi, J. B. Echouffo-Tcheugui, *et al.*, Understanding Contemporary Use of Thiazolidinediones, *Circ.: Heart Failure*, 2019, **12**(6), e005855.
126. M. Matta, A. Pavy-Le Traon, S. Perez-Lloret, *et al.*, Predictors of Cardiovascular Autonomic Neuropathy Onset and Progression in a Cohort of Type 1 Diabetic Patients, *J. Diabetes Res.*, 2018, **2018**, 8.
127. A. Gedebjerg, T. P. Almdal, K. Berencsi, *et al.*, Prevalence of micro- and macrovascular diabetes complications at time of type 2 diabetes diagnosis and associated clinical characteristics: A cross-sectional baseline study of 6958 patients in the Danish DD2 cohort, *J. Diabetes Complicat.*, 2018, **32**(1), 34–40.
128. S. Park, H.-J. Kang, J.-H. Jeon, M.-J. Kim and I.-K. Lee, Recent advances in the pathogenesis of microvascular complications in diabetes, *Arch. Pharmacal Res.*, 2019, **42**(3), 252–262.

129. K. Rett and U. Gottwald-Hostalek, Understanding prediabetes: definition, prevalence, burden and treatment options for an emerging disease, *Curr. Med. Res. Opin.*, 2019, 1–6.
130. A. Avogaro and G. P. Fadini, Microvascular complications in diabetes: A growing concern for cardiologists, *Int. J. Cardiol.*, 2019, **291**, 29–335.
131. A. Masuch, N. Friedrich, J. Roth, M. Nauck, U. A. Müller and A. Petersmann, Preventing misdiagnosis of diabetes in the elderly: age-dependent HbA1c reference intervals derived from two population-based study cohorts, *BMC Endocr. Disord.*, 2019, **19**(1), 20.
132. M. Kameyama, S. Takeuchi and S. Ishii, Steady-state relationship between average glucose, HbA1c and RBC lifespan, *J. Theor. Biol.*, 2018, **447**, 111–117.
133. R. R. Henry, J. P. Frias, B. Walsh, *et al.*, Improved glycemic control with minimal systemic metformin exposure: Effects of Metformin Delayed-Release (Metformin DR) targeting the lower bowel over 16 weeks in a randomized trial in subjects with type 2 diabetes, *PLoS One*, 2018, **13**(9), e0203946.
134. C. S. Birch and G. A. Bonwick, Ensuring the future of functional foods, *Int. J. Food Sci. Technol.*, 2019, **54**(5), 1467–1485.
135. A. F. Aydın, C. Küçükgergin, J. Çoban, *et al.*, Carnosine prevents testicular oxidative stress and advanced glycation end product formation in D-galactose-induced aged rats, *Andrologia*, 2018, **50**(3), e12939.
136. E. A. Guzmán, T. P. Pitts, M. C. Diaz and A. E. Wright, The marine natural product Scalarin inhibits the receptor for advanced glycation end products (RAGE) and autophagy in the PANC-1 and MIA PaCa-2 pancreatic cancer cell lines, *Invest. New Drugs*, 2019, **37**(2), 262–270.
137. W. Yu, X. Hu and M. Wang, Pterostilbene inhibited advanced glycation end products (AGEs)-induced oxidative stress and inflammation by regulation of RAGE/MAPK/NF-κB in RAW264.7 cells, *J. Funct. Foods*, 2018, **40**, 272–279.
138. P. Jha and A. Ali, Bionanoparticles in the Treatment of Glycation-Induced Secondary Complications of Diabetes, *Nanomaterials and Plant Potential*, 2019, pp. 299–316.
139. B. Sowndhar Rajan, S. Manivasagam, S. Dhanusu, *et al.*, Diet with high content of advanced glycation end products induces systemic inflammation and weight gain in experimental mice: Protective role of curcumin and gallic acid, *Food Chem. Toxicol.*, 2018, **114**, 237–245.
140. P. Zhang, Y. Li, R. Guo and W. Zang, Salidroside Protects Against Advanced Glycation End Products-Induced Vascular Endothelial Dysfunction, *Med. Sci. Monit.*, 2018, **24**, 2420–2428.
141. S. B. Agawane, V. S. Gupta, M. J. Kulkarni, A. K. Bhattacharya and S. S. Koratkar, Chemo-biological evaluation of antidiabetic activity of Mentha arvensis L. and it's role in inhibition of advanced glycation end products, *J. Ayurveda Integr. Med.*, 2018, DOI: 10.1016/j.jaim.2017.07.003.

142. A. Aljohi, S. Matou-Nasri, D. Liu, N. Al-Khafaji, M. Slevin and N. Ahmed, Momordica charantia extracts protect against inhibition of endothelial angiogenesis by advanced glycation endproducts in vitro, *Food Funct.*, 2018, **9**(11), 5728–5739.
143. L. Zhang, C.-j. Zhang, Z.-c. Tu, *et al.*, Nelumbo nucifera leaf extracts inhibit the formation of advanced glycation end-products and mechanism revealed by Nano LC-Orbitrap-MS/MS, *J. Funct. Foods*, 2018, **42**, 254–261.
144. G. Karan Rajpurohit, A. Roy and L. Thangavelu, Inhibition of Advanced Glycation End-Product Formation by Lutein from Tagetes erecta, *Pharmacogn. J.*, 2018, **10**(4), 734–737.
145. A. Schinkovitz, P. Le Pogam, S. Derbré, *et al.*, Secondary metabolites from lichen as potent inhibitors of advanced glycation end products and vasodilative agents, *Fitoterapia*, 2018, **131**, 182–188.
146. K. M. Glynn, P. Anderson, D. J. Fast, J. Koedam, J. F. Rebhun and R. A. Velliquette, Gromwell (Lithospermum erythrorhizon) root extract protects against glycation and related inflammatory and oxidative stress while offering UV absorption capability, *Exp. Dermatol.*, 2018, **27**(9), 1043–1047.
147. J. Uribarri, W. Cai, M. Peppa, *et al.*, Circulating glycotoxins and dietary advanced glycation endproducts: two links to inflammatory response, oxidative stress, and aging, *J. Gerontol., Ser. A*, 2007, **62**(4), 427–433.
148. B. Fernandez-Gomez, C. Nitride, M. Ullate, G. Mamone, P. Ferranti and M. D. del Castillo, Inhibitors of advanced glycation end products from coffee bean roasting by-product, *Eur. Food Res. Technol.*, 2018, **244**(6), 1101–1110.
149. L. Crascì, M. R. Lauro, G. Puglisi and A. Panico, Natural antioxidant polyphenols on inflammation management: Anti-glycation activity *vs* metalloproteinases inhibition, *Crit. Rev. Food Sci. Nutr.*, 2018, **58**(6), 893–904.
150. Q. Zhou, J. Gong and M. Wang, Phloretin and its methylglyoxal adduct: Implications against advanced glycation end products-induced inflammation in endothelial cells, *Food Chem. Toxicol.*, 2019, **129**, 291–300.
151. S. Mohamed, Herbs and Spices in Aging, in *Aging: Oxidative Stress and Dietary Antioxidants*, ed. V. Preedy, Elsevier, Oxford, UK, 2014, pp. 99–107.
152. X. Li, G.-J. Liu, W. Zhang, *et al.*, Novel Flavoalkaloids from White Tea with Inhibitory Activity against the Formation of Advanced Glycation End Products, *J. Agric. Food Chem.*, 2018, **66**(18), 4621–4629.
153. S. Gholipour, R. D. E. Sewell, Z. Lorigooini and M. Rafieian-Kopaei, Medicinal Plants and Atherosclerosis: A Review on Molecular Aspects, *Curr. Pharm. Des.*, 2018, **24**(26), 3123–3131.
154. S. K. Bharti, S. Krishnan, A. Kumar and A. Kumar, Antidiabetic phytoconstituents and their mode of action on metabolic pathways, *Ther. Adv. Endocrinol. Metab.*, 2018, **9**(3), 81–100.

155. J. Naveen and V. Baskaran, Antidiabetic plant-derived nutraceuticals: a critical review, *Eur. J. Nutr.*, 2018, **57**(4), 1275–1299.
156. D.-G. Han, S.-S. Cho, J.-H. Kwak and I.-S. Yoon, Medicinal plants and phytochemicals for diabetes mellitus: pharmacokinetic characteristics and herb-drug interactions, *J. Pharm. Invest.*, 2019, DOI: 10.1007/s40005-019-00440-4.
157. J. Tata and V. Jaitak, A Review on Molecular Mechanism of Flavonoids as Antidiabetic Agents, *Mini-Rev. Med. Chem.*, 2019, **19**(9), 762–786.
158. K. Bhaskarachary and A. K. R. Joshi, Chapter 11-Natural Bioactive Molecules With Antidiabetic Attributes: Insights Into Structure–Activity Relationships, in *Studies in Natural Products Chemistry*, ed. R. Atta ur, Elsevier, 2018, vol. 57, pp. 353–388.
159. A. K. Singh, V. Raj, A. K. Keshari, *et al.*, Isolated mangiferin and naringenin exert antidiabetic effect *via* PPARγ/GLUT4 dual agonistic action with strong metabolic regulation, *Chem.-Biol. Interact.*, 2018, **280**, 33–44.
160. K. C. Chinsembu, Diabetes mellitus and nature's pharmacy of putative antidiabetic plants, *J. Herb. Med.*, 2019, **15**, 100230.
161. M. Bilal, M. S. Iqbal, S. B. Shah, T. Rasheed and H. M. N. Iqbal, Diabetic Complications and Insight into Antidiabetic Potentialities of Ethno-Medicinal Plants: A Review, *Recent Pat. Inflammation Allergy Drug Discovery*, 2018, **12**(1), 7–23.
162. L. Porskjær Christensen and R. Bahij El-Houri, Development of an *In Vitro* Screening Platform for the Identification of Partial PPARγ Agonists as a Source for Antidiabetic Lead Compounds, *Molecules*, 2018, **23**(10), 2431.
163. J. Chaudhuri, Y. Bains, S. Guha, *et al.*, The Role of Advanced Glycation End Products in Aging and Metabolic Diseases: Bridging Association and Causality, *Cell Metab.*, 2018, **28**(3), 337–352.
164. J. Koska, A. Saremi, S. Howell, *et al.*, Advanced Glycation End Products, Oxidation Products, and Incident Cardiovascular Events in Patients With Type 2 Diabetes, *Diabetes Care*, 2018, **41**(3), 570–576.
165. M. Rasool, A. Malik, T. T. Butt, *et al.*, Implications of advanced oxidation protein products (AOPPs), advanced glycation end products (AGEs) and other biomarkers in the development of cardiovascular diseases, *Saudi J. Biol. Sci.*, 2019, **26**(2), 334–339.
166. W. Qu, C. Nie, J. Zhao, *et al.*, Microbiome–Metabolomics Analysis of the Impacts of Long-Term Dietary Advanced-Glycation-End-Product Consumption on C57BL/6 Mouse Fecal Microbiota and Metabolites, *J. Agric. Food Chem.*, 2018, **66**(33), 8864–8875.
167. K. Han, W. Jin, Z. Mao, *et al.*, Microbiome and butyrate production are altered in the gut of rats fed a glycated fish protein diet, *J. Funct. Foods*, 2018, **47**, 423–433.
168. M. Snelson and M. T. Coughlan, Dietary Advanced Glycation End Products: Digestion, Metabolism and Modulation of Gut Microbial Ecology, *Nutrients*, 2019, **11**(2), 215.

169. K. M. Zinöcker and A. I. Lindseth, The Western Diet–Microbiome-Host Interaction and Its Role in Metabolic Disease, *Nutrients*, 2018, **10**(3), 365, DOI: 10.3390/nu10030365.
170. L. Vitetta, Probiotics Can Break the Toxic Relationship Between the Intestinal Microbiome and the Kidney, *Dig. Dis. Sci.*, 2019, **64**(2), 297–299.
171. A. G. Casso, R. A. Gioscia-Ryan, Z. J. Sapinsley, *et al.*, Supplementation with the Gut Microbiome-Derived Metabolite Trimethylamine N-Oxide Induces Aortic Stiffening in Young Mice, *FASEB J.*, 2019, **33**(suppl. 1), 530.533.
172. M. Murakami and T. Hirano, The molecular mechanisms of chronic inflammation development, *Front. Immunol.*, 2012, **3**(323), DOI: 10.3389/fimmu.2012.00323.
173. N. M. J. Hanssen, C. D. A. Stehouwer and C. G. Schalkwijk, Methylglyoxal stress, the glyoxalase system, and diabetic chronic kidney disease, *Curr. Opin. Nephrol. Hypertens.*, 2019, **28**(1), 26–33.
174. S. Trellu, A. Courties, S. Jaisson, *et al.*, Impairment of glyoxalase-1, an advanced glycation end-product detoxifying enzyme, induced by inflammation in age-related osteoarthritis, *Arthritis Res. Ther.*, 2019, **21**(1), 18.
175. Z. Wang, J. Zhang, L. Chen, J. Li, H. Zhang and X. Guo, Glycine Suppresses AGE/RAGE Signaling Pathway and Subsequent Oxidative Stress by Restoring Glo1 Function in the Aorta of Diabetic Rats and in HUVECs, *Oxid. Med. Cell. Longevity*, 2019, **2019**, 14.
176. A. Scirè, L. Cianfruglia, C. Minnelli, *et al.*, Glutathione compartmentalization and its role in glutathionylation and other regulatory processes of cellular pathways, *BioFactors*, 2019, **45**(2), 152–168.
177. K. Prasad, AGE–RAGE stress: a changing landscape in pathology and treatment of Alzheimer's disease, *Mol. Cell. Biochem.*, 2019, **459**, 95–112.

Subject Index